Named in remembrance of
the onetime *Antioch Review* editor
and longtime Bay Area resident,

the Lawrence Grauman, Jr. Fund

supports books that address
a wide range of human rights,
free speech, and social justice issues.

The publisher and the University of California Press Foundation gratefully acknowledge the generous support of the Lawrence Grauman, Jr. Fund.

Your Brain on Altruism

THE POWER OF CONNECTION AND COMMUNITY DURING TIMES OF CRISIS

Nicole Karlis

UNIVERSITY OF CALIFORNIA PRESS

University of California Press
Oakland, California

© 2025 by Nicole Karlis

Cataloging-in-Publication data is on file at the Library of Congress.

ISBN 978-0-520-39759-0 (cloth : alk. paper)
ISBN 978-0-520-39760-6 (pbk. : alk. paper)
ISBN 978-0-520-39761-3 (ebook)

Manufactured in the United States of America

33 32 31 30 29 28 27 26 25 24
10 9 8 7 6 5 4 3 2 1

Contents

Acknowledgments *vii*
Prologue *ix*

I · Caring in Crisis

1. The Disaster Effect *3*
2. Bounded Solidarity *14*
3. The Loneliness Crisis *32*
4. The Coregulation Fix *49*

II · The Alchemy of Altruism

5. The Kindness Intervention *63*
6. Better Together *82*
7. The Mālama Mindset *102*
8. Volunteering for All *116*
9. Why Mattering Matters *126*

III · Sustaining Systemic Caring

10 The Power of Empathy *149*

11 Nature's Fire Extinguisher *158*

12 When You Can't Give, Witness *167*

   Epilogue *185*

   Notes *191*
   Bibliography *205*
   Index *219*

Acknowledgments

I would like to express my deepest gratitude to my family for supporting me during the process of writing this book, which started nearly a decade ago.

To my mom, Karen, who has given so much to me in her life. Thank you for telling me to never give up and caring for my daughter on my writing days. To my spouse, Ken, who took on extra work to support our family during my book leave. You have always believed in this book. Thank you for always being my first reader and editor. I love you. Thank you to Renata, who also helped care for my daughter as I wrote this book.

To my daughter, Frances, who opened my heart in so many ways and made me a kinder person in my everyday life. May this book inspire you and your generation to care for each other in the future.

To my extended family and friends, thank you from the bottom of my heart for your unwavering support for this book. To the Elevate Company in Benicia, thank you for being such a welcoming community and keeping me sane during my first year postpartum while writing a book.

To my editors and colleagues at *Salon*, thank you for your support and for giving me the time off to work on my book. This book wouldn't have been possible without our professional leave.

A special thanks to Naomi Schneider, my editor, who believed in this book and took it on as a project. None of this would have been possible without you taking a chance on me. Thank you to the team at UC Press, who approached this book with such intellect and kindness. And thank you to everyone who took the time to contribute to the book before it landed at UC Press. To my former agent, Elias Altman, thank you for championing this project for as long as you did. Thank you to Jennifer Dickinson for being my writing coach in the early days and always supporting this project. Thank you to all the beta readers for taking the time to read and comment on early drafts. I know your time is limited and it truly meant the world to me that you read my book and gave me your honest feedback. I took every comment into consideration and was humbled by the way you took the time to improve it.

My sincerest appreciation goes to my sources, who took the time to share their stories and research with me. You made this book such fun to write. I will never forget our in-person meetings and phone calls. You truly are the best and I'm so impressed by everything you're all working on. It was my honor to include your work in this book.

And finally, thank you to the readers who buy this book and spend time with it. I'm so grateful.

Prologue

If paradise now arises in hell, it's because in the suspension of the usual order and the failure of most systems, we are free to live and act another way.

REBECCA SOLNIT

A warm wind blew from the east under a swirling pink sorbet-colored sky. Napa Valley tourists clinked their wine glasses and toasted to another day in paradise. A gust of nature's breath was not entirely out of the ordinary for the area. Depending on the weather on the Pacific coast, and the season, California's wine country could get a bit breezy. Plus, the month of October is known for the so-called Diablo winds, the Bay Area's version of the Santa Anas near Los Angeles. At first, some may have embraced the moment to cool off, a respite for the typically hot and dry region during that time of year. But the ominous wind would soon transform a happy haven into a blazing inferno.

At 9:43 p.m. on October 8, 2017, the wind accelerated, dragging down an electrical wire, striking California's golden shrubbery like a dry match, and exploding in fire near a street called Tubbs Lane in Calistoga.[1] This event, later called the Tubbs Fire, turned much of

Napa County into food for a hungry wildfire. As firefighters flocked to the site, they reported more downed wires, exploding transformers, and an uncontrollable wind—all factors stacking up against any chance they had of containing the blaze. Embers flew miles away from the initial site, sparking more fires across the region.

Michael Rupprecht, who was a local junior in high school at the time, got out of school early the next day as the imminent danger of the wildfire became increasingly clear. At first, his family's home was safe. They weren't located in an evacuation zone. But physical safety didn't equate to normalcy. Rupprecht was wading through an unfamiliar zone of uncertainty as PG&E shut off gas lines in waves, worsening cell phone reception. As Rupprecht recalled in an interview with me five years later in Napa, the moment felt apocalyptic. Medical centers were being evacuated.[2] A primary school in Santa Rosa burned down.[3] Homes and buildings crumbled to ash. Smoke billowed over his hometown in an extraordinary way, blocking out the usual bright blue sky. Eventually, everything went dark. "And we were like, 'What do we do?'" he told me.

Due to the health hazards of being outside, Rupprecht and his friends knew that taking shelter inside was the safest option. Yet they unanimously felt like they didn't want to just sit around. In fact, doing nothing felt physically, mentally, and emotionally intolerable to them. Together, they hopped into his SUV and started driving around to evacuation centers, offering all that they had—their hands and functioning transportation—to help. At first, people declined their offers. Finally, an evacuation center said they needed assistance delivering supplies to and from another evacuation center. Rupprecht and his friends were elated.

Obviously, something bad was happening, he told me, looking at me with so much sincerity. It wasn't a moment that warranted feeling good or happy. Rupprecht and his friends were living through the middle of the most destructive wildfire in California's history.[4] But for the first time in a while he felt like he was part of something, like he belonged. He felt like he had a sense of purpose that he didn't experience in his everyday life. He felt connected to his friends in a different way than he usually did. It just happened to be a once-in-a-lifetime natural disaster. "I just remembered as grim as the situation was, I remember feeling like I was part of a tribe," Rupprecht told me, almost as if he was revealing a secret. "It was me and my buddies, and we were doing something for our fellow community members."

For a couple of days, Rupprecht and his friends kept volunteering to deliver supplies and donations to and from the evacuation centers. Over the next couple of weeks, the Tubbs Fire became one of dozens that were part of a massive "Northern California firestorm," which wouldn't be contained for another 23 days. Eventually, the deadly flames waned. But Rupprecht and his friends' altruistic efforts kept going. "We ended up with more fundraisers, new service projects, really small stuff, to keep it going," he said. "And we were all just noticeably happier."

I *Caring in Crisis*

1   *The Disaster Effect*

In California, wildfires are expected.[1] I've lived through a couple myself while living in the San Francisco Bay Area for over a decade. The smoke. The orange skies. I've always found it emotionally hard to drive through a forest that's been torched by fire. The charred branches that bend like melted metal and the black and gray palette are a difficult sight to stomach.

But what's often forgotten in the narrative of California's wildfires is that much of the state's ecosystems are fire dependent or fire adapted, meaning they rely on the destruction of fires to create new growth that the forest needs to thrive. Wildfires burn the buildup of organic debris of the ecosystem's vegetation composition. A wildfire can improve biological diversity by releasing new nutrients into the soil. After the flames have licked a tree's branches down to frightening levels of fragility, where survival and life look hopeless, new growth from plants such as ferns help reignite the process of restoration.[2]

Ferns could have evolved to have proverbial thorns, to be less inviting, to be less helpful to the forest as a whole, after a wildfire. They could have adapted to look out only for themselves. But that's not what nature intended or needed. Instead, their roots are tucked away underground, which allows them to protect themselves from fire. Once they survive, their first order of business is to make the ecosystem habitable for everything else. New plants keep the cycle of life going in the forest by stabilizing the soil. And the presence of new growth attracts wildlife like squirrels and mice, which will scavenge for the seeds that were dropped during the wildfire. Slowly the next generation of the forest's ecosystem is enlivened. This is partly why Indigenous people used to rely on so-called cultural burns as a form of land management.[3] In fact, the lack of prescribed burns is part of the reason why California's wildfires have become so dangerous today.

I don't want to anthropomorphize the resilience of new growth too much, but the fern does seem to be a fitting metaphor for the innate altruistic drive that humans feel in the wake of a natural disaster—like what Michael Rupprecht and his friends experienced after the Tubbs Fire, as described in the prologue. During the COVID-19 pandemic, Katherine May's book *Wintering* became an anthem of sorts.[4] In a society that promotes the idea that humans can and should maintain a constant state of happiness and productivity, May argued the natural rhythm of life isn't one long summer party. Instead, it's more like the cycle of a forest. There are times when life flourishes, and there are times when wildfires burn most of it to the ground. Humans are frequently forced to winter due to external circumstances. Despite the cold and isolating nature of the season, May wrote that winters "are our social glue." They bring people closer together. When we're faced with the rigid coldness of life, it's only natural for us to huddle together. Whether it's an unforeseen wildfire, a death, or depression, it's not winter itself that defines us but rather how we winter and how we emerge. "When everything is broken, everything is also up for grabs," she wrote. If done right, May wrote, we can leave winter wearing ferent sweater.

When a disaster like a wildfire strikes, those affected in the community often feel a unique sense of togetherness, an undeniable urge to help each other that they don't experience in everyday life. The acts of altruism,[5] which are when we act to improve someone else's well-being, witnessed or experienced amid such a tragedy are engraved in their minds forever. In return, recollections of the disaster aren't only defined by loss and tragedy, but by what was gained through both giving and receiving. And it's not just natural disasters that can have this effect, but any crisis that suddenly strikes and snaps people back into the moment to remind them of what matters the most. As Rupprecht alluded to, while his community literally burned to the ground, he never felt closer and more connected to his friends and neighbors. He also had never felt such a strong desire to help strangers before. Never did he think it would come at a time when he was most vulnerable.

In *A Paradise Built in Hell: The Extraordinary Communities That Arise in Disaster*, Rebecca Solnit investigated the unparalleled acts of altruism and generosity that arise in the wake of a disaster.[6] "Horrible in itself, disaster is sometimes a door back into paradise," she wrote. "The paradise at least in which we are who we hope to be, do the work we desire, and are each our sister's and brother's keeper." It's not a coincidence that although Rupprecht had friends who lost their homes and his entire community was devastated by the wildfires, when he reflects on that time it's almost as if he misses it. The phenomenon is comparable to how some veterans miss war, as Sebastian Junger explored in his book *Tribe: On Homecoming and Belonging*. We know that an outpouring of support is critical for helping communities and individuals recover in the wake of disaster. But acts of altruism are critical to our own wellbeing and survival. They are critical to our health. Yet more often than not, once the crisis passes and life returns to normal, the collective sense of looking out for one another and helping each other fades.

It first occurred to me over a decade ago that there could be health benefits to helping others. In the middle of my own quarter-life crisis, I found it hard to alleviate symptoms of loneliness. Half the time, my high-functioning anxiety felt manageable. My dentist helped me out. He gave me a retainer to keep my teeth from grinding like heavy machinery at night. In an attempt to connect to something bigger than myself, I went to yoga school in India. I certainly benefited from the mandated healthier lifestyle. A menu full of vegan food. Yoga and meditation classes every day for multiple hours. I found comfort in the new friendships I made. Still, there were moments I felt riddled with anxiety about the future. Plus, I wondered how sustainable it all was to escape from my everyday life back home.

One day, one of my yoga teachers kindly offered to take me to see his Vedic astrologer, whom he relied on to make serious life decisions, like who to marry or what job to take next. The astrologer sat behind an old, but functional, computer. I provided him with three simple details about myself: my birthday, time of birth, and location of birth. These three data points about myself managed to tell him that my "monkey mind," as he called it, was keeping me from being fulfilled in life. "You overthink too much," he repeated. In all seriousness, I asked him, "What can I do to fix it?" He quietly pondered, typed on his keyboard, and printed out a prescription. It wasn't for Zoloft or Wellbutrin. It was a prescription to do two acts of kindness. While I now see how my trip could be viewed as an example of the appropriation of "enlightenment culture"—a White middle-class woman from America engaging in Eastern practices—I did leave with one of the greatest lessons I've learned in my life: sometimes it's better for our health and happiness to focus on helping others than helping ourselves. While I certainly didn't travel to India to learn this, it clicked that it was the complete antithesis of what the self-care industrial complex tried to sell me back home—do anything, consume anything, to feel better.

In this book, I wanted to explore a hopeful possibility: What if this collective sense of caring for each other that is so strong in the wake of a disaster didn't fade? What if we maintained that desire to give more to others, rather than be solely focused on ourselves, in times when there isn't a crisis? Might that make us more resilient, help us have better health, physically and mentally, and feel more fulfilled in life, and less lonely? Could a culture that prioritizes caring for each other be what is needed for us to be better prepared to face the next crisis as a society? Could it be the way out of an invisible crisis that is affecting both our individual and societal health, an epidemic of loneliness?

To find answers, as a health and science journalist, I set out to investigate the health benefits of altruism, rooted in new science, and how doing good is good (if not better) for the one giving. I also wanted to learn how this collective sense of caring, if maintained, could help us survive all the disasters we are bound to face in the future. To do this, I interviewed neuroscientists, doctors, sociologists, psychologists, and disaster studies professors. I interviewed people whose lives changed through being more generous and people who got through their darkest times by being kinder. I learned that altruism has the power to provide us with a more

The Disaster Effect [7]

satisfying and lasting type of fulfillment that can possibly give us the gift of living longer. I learned that in order to benefit from giving, you don't have to give in some grandiose way. A culture of caring can start with something as simple as a shift in one's mindset. I learned that sometimes we are in a season of giving. At other times, we are in a season of receiving. We all benefit as witnesses of moral beauty.

My research led me to ask some fundamental questions. What if modern-day self-care, which is intended to prolong our survival, has become yet another form of individualism that harms us? Why is volunteering one's time so difficult for so many? What if acts of kindness are the antidote to the loneliness crisis? And if the answers are what I suspect they are, then why can it be so hard to maintain the state of altruism we experience amid catastrophe?

Throughout my reporting, I was haunted by the same questions repeating over and over again in my head: If we know helping others is good for us and can have truly profound effects on our individual and societal health, why can it feel so hard to be kind? Why is there burnout among caregivers in the United States? Why is kindness something that feels easier to embrace in the wake of a crisis and not in more "normal" times? Again and again researchers told me we live in a culture that's built on fear. And in order to prioritize caring for others, the systems that have been built on scarcity need to be burned to the ground. But I didn't want my book to end with an abstract answer that felt too big a task to undertake. Or one that was sprinkled with a bit of hopelessness. I wondered, could change come more organically and not require a crisis?

To find more answers, I interviewed people who are doing what they can to disrupt the systems to the best of their capabilities today. One is a former entrepreneur who is trying to start a social prescribing movement in the United States. Could we live in a country where doctors prescribe the act of volunteering to patients? It might shock you to know it's happening in the U.S. right now. I interviewed the leaders behind a program at Harvard University whose mission is to create a more caring and just world by flipping the script on academic success in early education and building empathy from a young age. I interviewed the leaders behind a government-led initiative in California to make volunteering more accessible to everyone in the state.

In the first part of this book, "Caring in Crisis," I focus on the phenomenon that Rupprecht and his friends experienced. Why is it that people come together in remarkable ways after a crisis with a deep desire to help others, even when they themselves are in danger? And what are the health benefits, if any, of being altruistic when you're in the wake of a disaster? In this part, I also explore a current crisis that I believe doesn't get enough air time, but could be one of the main reasons why it's so hard to cultivate a culture of caring in noncrisis times: the loneliness epidemic. In "Caring in Crisis," I argue if we don't make it through the loneliness crisis, the nuances of loneliness will keep us from coming together to face future threats that will impair both individual and societal health. The part ends with an explanation on why frequently suggested solutions, like socializing, aren't the answer, but rather how altruism can serve as a catalyst for the deeper connections we need to be resilient.

The second part, "The Alchemy of Altruism," focuses on the science of altruism and explores benefits through different scenarios—from random acts of kindness to group volunteering. At the same time, I weave in anecdotes and lessons from people who are prioritizing caring in our existing infrastructures today. In chapter 5, I explore the benefits of acts of kindness and report on a pediatrician's office in Ohio that's trying to make family kindness festivals a common occurrence in childhood. In chapter 6, I report on the latest research in the neuroscience of altruism and travel to Hawaii to learn about how state leaders are trying to get tourists to have a less extractive experience on their vacations, one where they don't only take from Hawaii but also give back. In chapter 7, I explain how volunteering regularly in groups is the gold standard in terms of reaping all the health benefits, which leads me to report on the California Service Corps in chapter 8. Finally, I report on why kindness and volunteering is so powerful for its participants, which all boils down to a sense of belonging and how the physical health benefits of having a sense of purpose can have an effect on a person's immune system and disease progression. I also start to explore how society can create a culture of what I call *systemic caring*.

In the third part of the book, "Sustaining Systemic Caring," I report on the ways in which our systems can be built to support kindness and caring, which starts with a focus on empathy at a young age. What if schools included regular lessons on how to be better listeners? In chapter 11, I explore the role oxytocin plays in feeling safe and enabling us to care. In the last chapter, I address the issue of caregiver burnout and propose that if we live in a society that cultivates caring, people can feel safe enough to meet each other where they are, no matter what season they're in—either giving, receiving, or witnessing caring.

• • •

August is a month of transition. A time of transformation. Harvest is near. The days are thick with tension between wanting to savor the last days of summer and managing the excitement of what's to come. But in Napa, there's an added complexity. Instead of transitioning from summer to autumn, the next three months are really just a second summer. There is no stereotypical fall in Napa County. Instead, it marks the peak of the wildfire season. As I drive through the vineyards, the sun's rays kiss my skin early in the morning. It's a warning that it's going to be a hot one. While it's been a quiet wildfire season as I write this in August 2023, I've lived in the Bay Area long enough to know that the charming golden hills are frequently thirsty for change. These postcard-perfect vistas are capable of bursting into flames at any given minute.

Beyond the picturesque vineyards, I drive up to Justin Siena High School and think about how Rupprecht and his friends must have been so disoriented from the smoke that took so much from their high school experience. But today isn't about the past, it's about the future. Rupprecht and his colleagues are hosting a fundraiser for the Napa Valley Education Fund as an initiative for his nonprofit called the Hero Foundation. Indeed, their altruistic efforts didn't fade after the Tubbs Fire—they were able to sustain them.

The mood, like the weather, is sunny. Music is playing over the loudspeakers at the high school's football stadium. Children are running around. A woman pushes her baby in a stroller. Players in neon green and pink jerseys chaotically dart across the green field like the edible ghosts in a Pac-Man game. Overseeing one of the competitions at the Heroes Cup as a volunteer referee is 21-year-old Carlo Bartalotti.

Bartalotti met Rupprecht at an event at a local FEMA center about a month after his family's house burned down during the 2017 Napa wildfires. Rupprecht, in his urgent mission to help others as his community burned to the ground, offered Bartalotti some support, "so I could repurchase some uniforms and textbooks that I needed to finish out the year," Bartalotti told me. Touched by Rupprecht's kindness, Bartalotti kept returning to events hosted by the Hero Foundation. Nearly eight months later, he joined the Hero Foundation as a volunteer. Since then, he's climbed the ranks as the foundation's senior marketing director. Bartalotti said receiving help from the Hero Foundation and his high school community was a fundamental part of his healing process, but maybe not for the reason one might think. Of course, the resources he received were helpful. But it was Bartalotti's introduction to volunteering himself that he really credits as life-changing.

"It changed my life for the better," Bartalotti said. "That's when I started really getting into service." His involvement in the Hero Foundation gave him "the tools" and "mood boost" to tackle the challenges that came with the pandemic a couple of years later, he added.

Despite going to college, Rupprecht and his friends never let go of their mission to help others. During the COVID-19 pandemic, the organization had the infrastructure in place to weather another crisis. In 2020, they held a fundraiser for health care workers on the front line of the pandemic. They volunteered together at a COVID-19 testing site in Napa County trying to stop the coronavirus from spreading. In 2021, they picked up 200 pounds of trash in Napa that otherwise would have found its way to the county's waterways. They've cleaned up parks, held blood drives and diaper drives, assembled care packages for the homeless, and even raised over $10,000 to send to troops in Ukraine.

"One day in the future, what if we can have an app where you can go on if you have time, and ask 'Where can I volunteer?'" Rupprecht told me.

His dreams are big and admirable.

Part of the draw of the Hero Foundation isn't just the opportunity to volunteer, but it's also the community. Sitting at a table at the Heroes Cup, three volunteers shared their stories about how they joined the group. For one 20-something named Nick, he attended one event months ago and felt so inspired by the kindness that he kept returning. By participating in the Hero Foundation and volunteering at events like that day's soccer tournament to raise money for a local education foundation to support local public schools, all three volunteers agreed that they felt more resilient to face future crises in their community. "It's a reminder that there's always somebody out there who is thinking about you and hoping that you're okay in those times," Nick said. "And we'll check in on you and that only happens when you come together and establish those connections going forward."

That's not to say that these goals and optimistic ways of thinking don't clash with everyday realities, especially for these college students. For example, if Nick had to either finish a school paper or attend a Hero Foundation event, he said he would think about what's best for him. But most of the time, he chooses to volunteer. "I always go for 'What does my mental health need?'" he said. More often than not, it's volunteering with his friends.

As I'm leaving the soccer tournament, a man asks me about my book. I tell him it's about altruism and its ability to build resilience. How can we, as a society, sustain this so-called disaster effect? He's intrigued, and contemplative. "There's so much hate in the world," he says. "It didn't used to be like this." He points to the teens I was just talking to and says, "Maybe they're on to something."

2 Bounded Solidarity

Cassandra Cotta's days were filled with planks, squats, and leg lifts.[1] The small-talk in between the five daily classes she led could sometimes be just as tiring as the workout, but she didn't mind having to interact with and be "on" for so many strangers. It came with the territory as a full-time Pilates instructor in 2019, when she'd surround herself with hundreds of people each day. Plus, not everyone was a stranger. As a transplant from New Jersey to New York City, she was delighted to have found a tight-knit group of friends at the studio she called her second home.

When the COVID-19 pandemic hit in March 2020, the 31-year-old experienced a massive shift, from spending her days around so many people in a workout space to being by herself in an apartment half its size. A week into the first lockdown, she lost her job, which not only put her in a vulnerable economic situation but also distanced her from the Pilates community, as she lost touch with her coworkers because she was unable to be in the same place as them. Despite her following recommended precautions from public health officials to keep the coronavirus at bay, it seemed as if nothing could protect her from the well of loneliness she was drowning in. She bought a cat. She went on runs. Yet, the loneliness persisted.

Four hours north on a suburban tree-lined street, Daniel Aldrich also struggled with loneliness despite being at home with his wife and two kids. However, as a professor at Northeastern University teaching disaster resilience, and a survivor of Hurricane Katrina, he knew what it would take to both survive and overcome his feelings of loneliness. He knew that despite all the tragedy around him, this could be an opportunity for growth and connection. Yes, that meant following mitigation strategies recommended by the Centers for Disease Control and Prevention (CDC), like adhering to Boston's lockdown measure, maintaining a six-foot distance from strangers, and wearing a mask. But he also knew that he'd have to be proactive in checking in on his neighbors and be readily available to help them when needed.

People coped differently with the stress and sudden life changes the pandemic caused. In some cases, the complete and sudden isolation, coupled with the external threat of an unknown deadly virus, led to an uptick in alcohol consumption and drug overdoses.[2] Other people expressed their anxiety by doing panic runs to the grocery store and fighting with each other over the last toilet paper roll. While all this was happening, Aldrich spent his time checking in on his neighbors and taking stock of what resources he had next door. If a neighbor didn't have a toilet paper roll, they knew that Aldrich was happy to give them one, and vice versa. Aldrich and his neighbors were creating their own mutual aid network, a reciprocal exchange of services and resources for mutual benefit.

"There is often a focus on individual preparedness in these times. 'Do you have enough water? Do you have a disaster kit?'" he said.[3] "Then the other approach is that the government is going to fix this, that there will be an investment from FEMA." But from his own research and experience, Aldrich has found that neither of

Bounded Solidarity

those responses is as successful as community-based ones, that is, when people who know each other build trust with one another and work together.

The concept of mutual aid—giving and receiving in the name of survival—is said to be as old as time. Philosopher Peter Kropotkin popularized the term *mutual aid* in his collection of essays titled *Mutual Aid: A Factor of Evolution*, which argued for cooperation, not competition. "Those species which best know how to combine, and to avoid competition, have the best chances of survival and of a further progressive development," he wrote. "They prosper, while the unsociable species decay."[4]

Aldrich said not only is the act of giving resources, which might seem counterintuitive in a time of crisis, key to survival, but genuine social connections are too. He cited the 1995 heat wave in Chicago as an example, where over 700 people died in five days.[5] "The people who passed away, it wasn't really a question of age or race or demographics, it was a question of social connectedness," Aldrich said. "People who were just vulnerable, like the elderly, didn't necessarily die because of that, it was when that vulnerability intersected with social isolation."

It used to be conventional wisdom that when humans faced a crisis it would inevitably bring out the worst in people. Hollywood movies are known to depict the wake of a crisis with people panicking and prioritizing individual interests over collective ones. For example, in the 2011 movie *Contagion*, Gwyneth Paltrow contracts a virus that starts in Hong Kong and kills over 20 percent of people who get infected. The movie is about the loss of social order as the virus turns into a worldwide pandemic. Like in real life when the COVID-19 pandemic hit, the people affected in the film are told to self-isolate to protect themselves, as there is no vaccine or treatment for the new virus. As a result, anarchy unfolds. But as researchers noted in a paper coincidentally published right before COVID-19 spread across the world, the breakdown in social order, the panic often depicted in these thriller films, doesn't align with reality. "Cinema is not reality, and elements of fear leading to panic and chaos are likely to persist in dramatic representations of disease, purely for entertainment value," the researchers wrote.[6] *Contagion* isn't the only movie to highlight a common narrative about crises in mainstream media. More recent apocalyptic TV drama series have depicted humans as innately selfish in the wake of a disaster, as in *The Last of Us* and *Station Eleven*. The every-person-for-themselves narrative depicted in the media makes people believe that will be our reality. There have been numerous news articles about billionaires building secret bunkers to prepare for future catastrophes.

But as we saw with COVID-19, people initially rallied together and risked their own health to help others.[7] Some people volunteered at food banks. Many people came together to find proper personal protective equipment for first responders. Some even started sewing their own masks to donate to health care professionals. Others stepped up to volunteer and grocery shop for their vulnerable neighbors or to disinfect public transportation.

Society's response to a disaster has intrigued researchers for decades. After World War II, the U.S. government was curious to know what would happen to the country if a wartime nuclear attack actually occurred. The researchers weren't so much interested in the physical damage to cities and towns, and the obvious adverse health consequences, but were more interested in how people would respond. Specifically, some questions the government had were *What types of people are susceptible to panic and what types can be counted on for leadership in an emergency? What aggressions and resentments are likely to emerge among victims of a disaster and how can these be prevented from disrupting the work of disaster control?*[8] To find out, the government commissioned the National Opinion Research Center (NORC) at the University of Chicago to conduct qualitative fieldwork on disasters. Clearly, the researchers couldn't simulate a wartime disaster, so they looked for a proxy, and they figured that a natural disaster, like an earthquake or a tornado, was just as socially disruptive. The findings helped shape the unexpected nuances of disaster studies, which notably came to light during the aftermath of Hurricane Katrina.

For Aldrich, panic wasn't a major theme of his experience surviving Hurricane Katrina, despite what many media reports depicted. In fact, being a survivor of Katrina led him to study resilience in the wake of disasters. He believes altruism is key to survival and disaster preparedness agencies should promote prosocial behaviors, such as volunteering and solidarity, in noncrisis times to better build resilience in society.

Jennifer Trivedi, an assistant professor of anthropology at the University of Delaware, who studied disaster recovery after Hurricane Katrina, agreed. "On the ground, it really boiled down to local people and local groups, and small groups and individuals kind of helping one another," she told me in a phone interview. "So a lot of those resources that worked kind of for the best got channeled through local groups and local people on the ground who kind of knew who needed what and who needed help in their community."

More recently, researchers have found that social connections were also critical to surviving one of humanity's worst tragedies, the Holocaust. In a study published in the *Proceedings of the National Academy of Sciences* (PNAS), researchers analyzed survivor testimonies to understand the link between social ties and vival.[9] They found that entering Auschwitz with a larger group potential friends likely provided people with a significant survival advantage. My grandfather, who was a Holocaust survivor, alluded to this in his own testimony that wasn't part of the study. He said there was unspoken "passive resistance" among prisoners. They looked out for each other. It helped them survive.

. . .

A little less than a century before the Napa wildfires, on December 6, 1917, in the waters off the coast of Halifax, Nova Scotia, a French cargo ship, the SS *Mont-Blanc*, collided with a Norwegian vessel, the SS *Imo*.[10] Accidents were common in busy seaports back then. But this one made headlines worldwide.

The problem was that the SS *Mont-Blanc* was a massive floating explosive, as it carried TNT, gun cotton, high explosives, and 3,000 tons of picric acid, which was used for the production of explosives and matches. When the collision happened, a fire started on the SS *Mont-Blanc*, unleashing a big cloud of smoke. People on shore curiously ran outside their homes to see what had happened. The crew shouted from their lifeboats: an explosion was imminent. Run. But the shouts couldn't save them. A few minutes later, the ship exploded and killed over 1,600 instantly. Hundreds of people who were watching the smoke from outside their homes were blinded. Windows shattered. Turned over gas stoves lit on fire, proceeding to burn down entire city blocks. Today, it remains one of the biggest nonnuclear explosions in history.

After the disaster, a man named Samuel Henry Prince started a PhD program in sociology at Columbia University, where his professor suggested that he use the Halifax explosion as the basis for his thesis on a theory he referred to as "collective behavior," describing how social change can be possible in the wake of a disaster. "Catastrophe becomes also the excitant for an unparalleled opening of the springs of generosity," he wrote. The thesis, "Catastrophe and Social Change," became the first systematic study of disaster and is still quoted and highly regarded by experts in the field.[11] And it became the first document to suggest that massive social change can emerge from a crisis.

What Prince was referring to has evolved into a concept called "bounded solidarity" today. Alejandro Portes, a prominent sociologist and emeritus professor at Princeton University, first introduced the term in a paper published in the *Annual Review of Sociology* in 1998.[12] "It is a source of social capital that is elicited by providing people feeling a sense of communality or loyalty with others in their own particular community," Portes told me in a phone interview. "It is 'bounded' because it is not in solidarity with everyone else in the world."

Bounded solidarity is frequently prompted by a disaster, crisis, or something that binds people together, which causes people to see themselves as a collective, not as individuals. As a result, this can create a different kind of solidarity from what exists in noncrisis times, which is why it can lead to extreme acts of altruism.

Unlike what Portes refers to as "value introjection" in his research, where people are motivated to be altruistic due to their holding specific religious values, bounded solidarity creates a unique sense of togetherness. Portes explained to me it's what encourages donors from a particular religious or ethnic group to contribute scholarships to students of the same religion or ethnicity. It's also what can bring immigrants closer together while facing marginalization. It's why for some low-income communities generosity is a survival tactic, not a luxury.

Like he said, it is also common to see bounded solidarity emerge in the wake of a natural disaster. It doesn't always happen, but most of the time it does. He even wrote a book called *City on the Edge: The Transformation of Miami*, noting the way bounded solidarity emerged after Hurricane Andrew in 1992. As a Category 5 storm, Hurricane Andrew caused $20 billion in property losses, left 160,000 people homeless, and destroyed or damaged 82,000 businesses.[13] Portes found it intriguing that the hurricane happened at a time when there were social and racial divides in the city of Miami. Despite these differences, Portes said, people who were in conflict with each other were able to come together and quickly put their differences aside in the wake of the hurricane. "For several months, their conflicts that existed, their big legal issues in the community, were attenuated," Portes told me.

In his book's postscript, Portes cautiously speculated that it might have been possible for Miami to sustain its bounded solidarity, that despite the differences the community faced in noncrisis times, a more "profound convergence" was possible. He believed that the identity of the area could be built on the solidarity of overcoming Hurricane Andrew and that it would serve as a point of connection for survivors and future inhabitants. The bounded solidarity Miami experienced would, in Portes's words to me, transform into "durable solidarity."

Considering that the sea level along Florida is expected to rise over the next century,[14] putting cities like Miami at risk of flooding, "durable" solidarity could hypothetically make the city more resilient to face future climate-related events. Unfortunately, Portes said, solidarity is no longer bound in Miami—at least around

Hurricane Andrew. "It lasted only about a year afterwards, and then things reverted to the usual patterns of divisions in terms of where people live, where they work and so on," Portes said. "So at present, not many people here in this metropolitan area remember what happened in 1992."

In the context of a crisis, bounded solidarity happens when there is a shock to the system that causes people to set aside their everyday pursuit of personal gain in a society that is built on scarcity. In Naomi Klein's book *The Shock Doctrine*, she explored how this can sometimes be an opportunity for the private sector to take advantage for their own interest. Author Rebecca Solnit took a more hopeful approach, analyzing how, as Prince believed, it can be a pathway to exceptional acts of altruism and some sort of social utopia. Despite the power of bounded solidarity in the moment, and the opportunity it presents, it sadly doesn't appear to be very resilient.

In unique cases, like the story of Rupprecht from Napa, it is to some extent. However, as we saw during the COVID-19 pandemic—and as Portes told me—the individual pursuit almost always returns for most people. Too often, bounded solidarity unbinds. "Most people in that situation, once the emergency ends, revert to their career and economic pursuits," Portes said. When I asked Portes why, he said that it is usually because the forces of individualism take priority. "Remember, we live in a capitalist society, right?" he said. "The basic, motivational drive under capitalism is the pursuit of money." It's not a coincidence that sociologists like Portes attribute the origins of bounded solidarity to German philosopher Karl Marx's vision for a socialist society and the idea that a society driven by money fuels class separation.

For decades, the majority of disaster researchers have focused on the response and recovery phases of a crisis. Adam Straub, a

PhD student in sociology at Oklahoma State University, was curious to look at how bounded solidarity operated from a preparedness perspective. For example, what happens before a disaster strikes? If a community already exhibited a version of bounded solidarity, if it didn't fade, did it make a community more resilient?

To find answers, Straub and his colleagues spent five years interviewing people in rural Oklahoma. As the researchers noted, Oklahoma is predisposed to a variety of climate change–related extreme events, such as severe thunderstorms, tornadoes, ice storms, and floods. Despite this, emergency services across the state have experienced huge budget cuts. What Straub and his colleagues found was that the survival of rural communities depended on what they referred to as the "Oklahoma spirit."

In their research, Straub and his colleagues heard multiple stories about how community members pooled their resources to help each other when public institutions failed to assist them. Since money to fund programs to be prepared has to be drawn from local tax revenue rather than state allocation, rural communities are often left vulnerable due to a lack of funding. As a result, Straub and his colleagues found that people in rural Oklahoma survived on altruism and reciprocity instead of funding.[15] Through coming together and volunteering at higher than average rates compared to their peers in more-resourced areas, their communities built systems of resilience. "They would all saddle up and go to help them because they understood that if it happened to them, that community would have done the same," Straub told me.

Straub said he personally found it fascinating how volunteers helped alleviate a lot of the deficiencies they had as a community despite being underresourced financially. For example, many of the towns didn't have the funds to staff emergency departments, so they would rely on people volunteering to be firefighters or be part of a search and rescue team. Many times, Straub said, these volunteers would donate their time in multiple sectors, since funding was not the only issue, as they struggled to find "manpower" as well. These volunteers were so motivated to help each other and be there for one another that they paid for various training courses themselves. Straub said this overlap raised the communities' "human capital," which is their knowledge of what to do in an emergency. Collectively, this mix of altruism and human capital, Straub said, made these communities extremely resilient and prepared to face future crises. "In terms of these crises, their disaster knowledge and acumen, if you want to call it that, will be higher than those in urban areas who don't have to do any of those things," Straub said. "I think in some ways, the average community member would be more capable and prepared than folks would be in an area that has dedicated personnel and services to those types of things."

Various communities appeared to have mutual aid networks with communities that they weren't directly helping and participating with on a regular basis, too. For example, instead of referring to a sheriff in another community, people would refer to said sheriff by that person's first name. "That would imply to me that it was like a more intimate, informal and close-knit kind of relationship," Straub said. "I would argue that as opposed to urban areas that have more funding and have agencies that are directly charged with and paid to do this on a full-time basis, they [Oklahomans] are more prepared." These communities appeared to have a better capability of working together and trusting one another.

That's not to say this type of bounded solidarity doesn't have its downside. "We found that because there was this very antipathetic or animosity towards the state and cities, more broadly speaking, that they became very insular and sealed off from accepting help from those folks," Straub told me. In this sense, Straub said, bounded solidarity can be isolating and far from the paradise that some argue can emerge out of a disaster. Straub also found it challenging to glorify these findings and point to them as a blueprint for disaster resilience because this kind of social utopia existed as a consequence of disenfranchisement. "Those prosocial behaviors don't necessarily require a disenfranchisement," Straub said. "I think people in general, kind of like in a Rousseau's social contract kind of way, deep down don't want to see other people suffer, and if they can help to do something about it, and they're able to, they will." But once people are no longer in peril, there is a decay. People's experiences go back to normal or a new normal after a disaster. "And so does social life, right?" Straub posited.

• • •

At first glance, an ant appears to be a tiny, vulnerable insect. Moving across a man-made pathway, a solo ant can be squashed in an instant by a human. For centuries, ants have caught the interest of scientists, especially biologists, but they are also a nuisance. I'm reminded of the latter after a rainy day when I witnessed an army of ants crawling across my kitchen countertops.

But I also have to remind myself that when we see ants above ground they aren't in their safe space. Each ant on earth belongs to a colony and lives in a series of underground chambers that are connected by small, complex tunnels.[16] These colonies exist within a caste system, which is determined by the nutrition each ant receives as larvae. Queen ants are the founders of the colonies.

After mating once, they are able to produce millions of eggs for years, up to 1,000 eggs a day. Each queen ant can live between 1 and 30 years. Depending on the species, worker ants tend to not live as long. This is because they are responsible for most of the work that goes on in a colony. As sterile females, they forage for food. Their modus operandi is to work hard to protect their colony, to care for it. When we see solo ants, their job is to risk their life for the benefit of their species, just like solo bees. Some ants even explode and kill themselves as the ultimate act of defense.[17]

Charles Darwin's natural selection theory posits that the way for a species to survive depends on how an organism adapts to their environment. Adaptation leads to survival when the genetic mutations that benefited an individual's survival are passed on through reproduction. For example, giraffes' long necks gave them an advantage by allowing them to feed on leaves that other species couldn't reach. The fact that ants were so seemingly altruistic, and that some worker ants evolved to be sterile and yet the species succeeded in reproducing, caused Darwin to pause. In fact, it created a paradox that threatened his theory of evolution. "He who was ready to sacrifice his life, as many a savage has been, rather than betray his comrades, would often leave no offspring to inherit his noble nature," Darwin wrote in *The Descent of Man*.[18]

"Survival of the fittest" is often misattributed to Darwin. It was actually coined by philosopher Herbert Spencer and has been used to justify social inequality, eugenics, and racism. More accurate interpretations of Darwin's work, like *The Descent of Man and Selection in Relation to Sex*, emphasize the fact that cooperation, not competition, is key to survival.

Nearly a century after Darwin, British biologist William D. Hamilton looked at the evolutionary underpinnings of social bonds.[19] This led to his theory on how a gene for helping others, even if it meant sacrificing one's own life, could be passed along. It was first referred to as "inclusive fitness" and later evolved into "kin selection."[20] Evolutionary biologist George Williams summed up the theory in his 1966 book *Adaptation and Natural Selection*.[21] "Simply stated, an individual who maximizes his friendships and minimizes his antagonists will have an evolutionary advantage, and selection should favor those characters that promote the optimization of personal relationships," he said.

The idea that an altruistic gene exists and is responsible for species survival is exhilarating. But as many have written before me, it's inaccurate to see eusocial insects as analogies for human beings. Humans are far more complex. "Let us not aspire to be like ants and bees," Katherine May wrote in *Wintering*. "We can draw enough wonder from their intricate systems of survival without modeling ourselves on them wholesale."

Maybe the lesson from ants isn't to be more like ants, but instead wonder why we live in a culture where "survival of the fittest" takes precedence over "survival of the kindest." Why is the default culture, when bounded solidarity fades, a dog-eat-dog world?

When I was pregnant with my daughter, I would sit alone in her nursery and get lost in my imagination, thinking about what life would be like when she arrived. The room was full of reminders of anticipation, like an old edition of Aesop's fables that my father-in-law read to my husband as a kid. This group of stories, thought to have been written by the Greek storyteller Aesop, were created to demonstrate moral lessons. One day, the fable *The Lion and the Mouse* caught my attention. In the story, a little mouse runs across a lion's nose while he is sleeping. Roused from his sleep, the lion angrily lays his paw on the mouse, threatening to kill him. But the mouse pleads for his life. "Spare me!" he begs. "Please let me go and someday I will surely repay you." The lion didn't believe that a tiny mouse, of all creatures, could ever help him. He was an apex predator and there was no benefit to him to let the mouse go. But he did. Sometime later, the lion was hunting prey in the forest and got caught in a hunter's net. Helplessly, he roared in one last attempt for someone to save him. From a distance, the mouse recognized his roar. He ran to the lion, gnawed on the rope, and ultimately set him free. "You laughed when I said I would repay you," the mouse said. "Now you see that even a Mouse can help a Lion."

The universal moral of this Aesop fable is that an act of kindness is never wasted. But if I take another look at it, I find a different takeaway. It's not only a moral lesson of kindness, but one about the nuances of survival. The lion, after all, is not an animal that is known for being particularly generous. A lion represents strength and power, stereotypical masculine energy. Modern humans fear the lion. We are told to stay away. They'll kill us with their knife-sharp carnassial teeth. We leave them alone out of fear. But in ancient times, lions frequently symbolized a complex narrative about destruction. In the Book of Revelation, which focuses on the forces of evil that have taken over the world, Jesus resurrects as the Lion of the Judah. Just as the lion represented the power of the sun to ancient Egyptians, the Lion of the Judah was seen as a destroyer of darkness that brought light back into the world. Is it possible that Aesop was trying to communicate that not only does an act of kindness never go unrewarded, but that there's strength in being kind? Or, put another way, that the lion's life depended on how he cared for others, not how much he acted out of his own self-interest?

For nearly a century, scientists have tried to prove the importance of altruism, cooperation, and kindness in human evolution. In one case study, researchers Felix Warneken and Michael Tomasello, at the Max Planck Institute for Evolutionary Anthropology, were curious to know if nonverbal toddlers had the capacity to help others.[22] In other words, is it possible that altruism is something we as humans are innately driven toward even from a young age? In one of their famous experiments, published in the journal *Science* in 2006, they found that when they dropped a clothespin and subtly signaled that they needed help, nearly every young child tried to provide assistance. The researchers performed the study in front of a group of chimpanzees who were raised by adults and found that their apparent instinct was to help, too. Previously, theorists had claimed that the tendency of human beings to cooperate with one another was not found in other animal species. "The current results demonstrate that even very young children have a natural tendency to help other persons solve their problems," the researchers concluded.

It's strange that something that seems to come naturally to both toddlers and, later, adults in the wake of a disaster is referred to as the "Honeymoon Phase" of a crisis. Experts believe people go through a series of psychological reactions as a disaster unfolds: Heroic, Honeymoon, Disillusionment, and Recovery.[23]

The Heroic Phase occurs at the time of the disaster and immediately after. It is when evacuations are happening, shelters are opening their doors, and the totality of the disaster is yet to be fully understood. The Honeymoon Phase begins a few days after the disaster and can last for several months. This is when altruism is at its peak. Neighborhood and community groups make it a priority to gather together to support each other. The Disillusionment

Phase is when survivors begin to get overwhelmed at the prospect of rebuilding their lives; this can last for several months or years. It's when bitterness, disappointment, and anger may come into the picture; it is also when major relief agencies (such as the American Red Cross) leave the disaster area. Sometimes, this can create a sense of abandonment among the survivors. The Recovery Phase is when the survivors are on their own to rebuild and move on from the disaster. A new sense of activism may occur during this time. What is unclear is how altruism, volunteerism, and random acts of kindness continue after the Recovery Phase and what possibilities they can yield if they are sustained.

When asked to define the difference between resilience and recovery after a disaster, Aldrich said resilience is "the ability of an individual, institution, or community to go through a shock and ideally transform." A baseline, or the lowest level of resilience, would simply mean being able to resume some kind of normalcy afterward. Aldrich pointed to Japan's Tohoku earthquake in 2011, which caused a tsunami. Rebuilding homes on the coast wouldn't necessarily be an example of resilience in the communities affected by the tsunami; rather, moving people uphill, building bigger seawalls, and "changing the system" to make future disasters less damaging would build resilience and lead to true transformation.

Take, for example, the United States as a whole after the COVID-19 pandemic. Aldrich said in order to measure recovery versus resilience, he'd ask businesses about the number of clients they had who have returned. Cities, he added, can measure how many cars are in the streets and how many kids are catching school buses—what is referred to as "mobility data." If these numbers match those from the predisaster time, that's a signature of societal resilience. Aldrich said population is another example of a tangible way that resilience can be assessed. When considered in resilience, American cities don't stack up very well. As Census Bureau found, the growth of what were considered to be the fastest growing cities in 2019 slowed down during the pandemic.[24]

As a society, America is perpetually stuck in recovery mode. After I spoke with Aldrich, it became clear to me that recovery is marked by stagnation. Resilience is marked by transformation. Crisis after crisis, from mass shootings to wildfires, we sometimes recover—but we rarely transform. Why?

3 The Loneliness Crisis

The HBO show *The Sopranos* begins with Tony, the family's patriarch, trying to explain to his therapist, Dr. Melfi, why he's having panic attacks.

The day he had a panic attack, he was thinking, "It's good to be in something from the ground floor." Admittedly, he's too late for that, but lately he's been getting a feeling that he came in at "the end" of something and that "the best is over." Dr. Melfi responds that she thinks many Americans feel the same way. Tony elaborates. He thinks about his father. He never reached "success" like Tony, but he says his father had it "better" in a lot of ways. That's because he had his people, they had their standards, and their pride. "Today, what do we got?" he asks.

This is perhaps the most analyzed scene from *The Sopranos*. While Tony is talking about the deterioration of the Mafia, the HBO show is a metaphor for the decline of America. Its cultural commentary focuses on the ruthlessness of capitalism and individualism, and the consequences of how extreme it all had become by the 1990s. As the show's creator David Chase once said, the intent behind *The Sopranos* was to create a show that depicted that "life in America had gotten so savage, selfish, that even a mob guy couldn't take it anymore." Even Tony Soprano, whose profession was centered around cheating the system and murdering people to make a profit, a man whose business was built on a "me first" attitude, felt isolated due to the constraints of late-stage capitalism. Even he craved, in some sense, the luxury to prioritize cooperation over competition. Instead, Tony found himself unable to trust even his closest confidantes.

Perhaps it's not a coincidence that a year after the premiere of *The Sopranos* in 1999, Robert Putnam's book *Bowling Alone: The Collapse and Revival of American Community* struck a chord with people, too. Based on his 1995 essay titled "Bowling Alone: America's Declining Social Capital," published in the *Journal of Democracy*,[1] the book garnered attention from major media outlets like the *New York Times* and National Public Radio in part because many Americans felt that their society was witnessing a decline in civic engagement and the consequences were palpable. As Putnam explained, voter turnout had experienced a major decline between 1960 and 1990. Fewer people were attending public meetings on town or school affairs. Labor union membership experienced a drastic decline between 1975 and 1985. The percentage of Americans socializing with their neighbors more than once a year had plummeted.

However, most peculiar to Putnam was the trend of American bowlers. In the essay, he explained how there were more Americans bowling in 1995 than ever before. In fact, the sport of bowling was on the rise. However, the number of people bowling in organized leagues had decreased by nearly 40 percent. More people were bowling, but they were doing it alone. He concluded that this not only threatened the existence of bowling lanes, but it had bigger implications for society. "The broader social significance, however, lies in the social interaction and even occasionally civic conversations over beer and pizza that solo bowlers forgo," he wrote. Life is easier, more enjoyable in a society where there's high social capital, Putnam argued, which can be built by bowling together, not alone.

Many researchers argue that social capital is a vital sign of a society's health and democracy;[2] high social capital indicates a healthy society, while low social capital suggests a society is sick. In order for social capital to exist and thrive, people must have high levels of trust and cooperation with each other. However, it's not just something that can be built and sustained in the wake of a disaster. Like a plant, it needs to be nurtured and maintained. In his essay, Putnam said that when people actively participate in civic activities, it builds a culture of reciprocity and trust, the key ingredients to sustaining social capital. The mere fact that fewer people were bowling together was a threat to American democracy.

Putnam didn't leave readers hanging as to why he believed fewer people were engaged in civic organizations. In his essay, he offered four trends that could explain why civic engagement was down. First, he said that more women had entered the workforce, suggesting that women made up most of the participants in civic engagement activities. Second, he said it could be an adverse effect of people moving around America more, and as a result uprooting their social lives and finding it difficult to plant social roots elsewhere. Third, it could be due to overall demographic changes, such as more divorces, fewer children, fewer marriages, and lower wages. "Each of these changes might account for some of the slackening of civic engagement, since married, middle-class parents are generally more socially involved than other people," Putnam said. But perhaps it's the final possible cause that would resonate the most today: the "technological transformation of leisure."

Putnam said that there was reason to believe, in the 1990s, that leisure time was being "privatized" or "individualized" by technology. He pointed to the rise of television as an example. In the 1960s, the television started to dictate how Americans spent their days and nights. "Is technology thus driving a wedge between our individual interests and our collective interests?" Putnam asked.

It wasn't all baggy jeans, the Spice Girls, Tamagotchis, and Bill Clinton in the 1990s; the decade also marked the dot-com boom. Since the 1990s, many of the trends that Putnam noted were on the decline are still eroding today. PTA memberships are barely hanging on, and since the pandemic, many chapters are no longer meeting in person.[3] In 2014, the U.S. saw its worst voter turnout for a midterm election in 72 years.[4] The percentage of U.S. workers in a labor union continues to decline. In 1983, 20.1 percent of American workers were union members. In 2023, that number had halved to 10 percent. In the 2017 book *One Nation after Trump*, the authors wrote that the decline of social and civic groups documented by Putnam was a contributing factor in the election of Donald Trump. "Many rallied to him out of a yearning for forms of community and solidarity that they sense have been lost," they wrote.[5]

Not only has this been detrimental to American democracy, but it has also contributed to another crisis: the loneliness epidemic. Before the COVID-19 pandemic, one in two adults in America reported experiencing loneliness.[6] The crisis peaked when Dr. Vivek Murthy, the United States' 21st surgeon general, released an 85-page advisory declaring loneliness a public health epidemic in America in 2023. In the advisory, Murthy noted that over the past few decades, Americans have become increasingly less socially connected. People are spending more time alone, especially young people, whom researchers say are spending less time with friends. "This decline is starkest for young people ages 15 to 24," the advisory stated.

· · ·

On January 9, 2007, Steve Jobs stood in front of an oversized projector screen at MacWorld to announce what Apple enthusiasts could expect to see from the tech conglomerate that year.[7] First, a wide-screen iPod with touch controls. Second, a "revolutionary mobile phone." Third, a "breakthrough internet communicator." Wearing his iconic black turtleneck, Jobs cycled through the three slides a couple of times and asked the crowd, "Are you getting it?" They weren't three separate devices, he emphasized. They were one, and he called it the "iPhone."

Previously, smartphones existed in the form of a Blackberry. They had plastic control buttons. Applications were few and far between, and they certainly couldn't be improved with frequent, personalized updates. To solve that, Jobs and his team built the iPhone, which he described as "magic for your hand." "We're going to get rid of all these buttons and make a giant screen," Jobs said. And you wouldn't need a stylus because you were born with 10 of your own.

More than 15 years later, our human styluses are very occupied these days. On average, Americans spend at least six hours on digital media per day. An estimated one in three adults report that they are "almost constantly" online.[8] As the magic of the iPhone settled, 10 years after Jobs excitedly cycled through slides, rousing the crowd, the creators of the iPhone expressed concern about how much time people spent on their screens—an unintended consequence they said they didn't expect. In an interview with journalist Brian Merchant, former Apple employee Bas Ording said one downside of the interface is that "now too many people are staring at their phones." "In terms of whether it's net positive or net negative, I don't think we know yet," Greg Christie, Apple's former vice president of human interface said in an interview. "I don't feel good about the distraction."[9]

The average time Americans spend on screens is seven hours and four minutes per day.[10] In 2019, a study found that Americans averaged more than five hours of free time each day.[11] The study sought to dispel the myth that Americans didn't have enough free time to exercise. Researchers found men generally had more leisure time than women. But instead of being physically active during that time, they were spending their free time on their phones. Like Putman speculated in 1995, our leisure time is still being influenced by technology.

Six months after Jobs announced the invention of the iPhone, it went to market and the public could have their own iPhones. This is the day Dr. Don Grant, a media psychologist and National Advisor of Healthy Device Management for Newport Healthcare, told me, "Everything changed." "June 29, 2007, was a game changer," he said. "We crossed the event horizon and it was the inflection point."

For 17 years, Grant has spent his time studying the potential impact of smartphone use behaviors on kids, tweens, and teens. One of his theories is that young people are vulnerable to the neurological and psychological effects of overuse and that society needs to help them take back control. He believes this is partly why we are seeing fewer in-person interactions among young people.

"Kids don't even know how to extend an organic bid for connection," he told me. "My theory is that June 29, 2007, that's when the world changed, because when the internet became portable, through the iPhone, then suddenly, all we want is that iPhone and they knew it."

• • •

Dr. Steve Cole, a genomics researcher at the University of California, Los Angeles, who studies the health effects of loneliness at a molecular level (you'll meet him again later in this book), told me the apps and technology that were meant to make everyday life activities easier are exacerbating the loneliness epidemic. As humans succumbed to that allure of everyday activities being co-opted by apps, humans gave away opportunities to feel connected and cared for by acquaintances and other people in our everyday experiences.

"It commodifies human beings and leads to a sort of creeping disrespect and contempt for the value of other human beings," he told me. One example he gave is going to a butcher to obtain meat. In having to get meat from the butcher, it was only natural that the consumer would strike up a conversation with the butcher. Despite their differences, which could have been political, they both needed something from each other. The consumer needed meat. The butcher needed customers. In this exchange to meet their respective needs, they'd get to know each other organically over time and gain trust in one another. If the butcher's kid was sick, the consumer would listen and offer support. The consumer would have an organic opportunity to show compassion and empathy. The butcher, in a fleeting moment, could feel cared for and connected to another person. Now, technology has commodified that type of exchange. Another way technology has contributed to the loneliness epidemic, Cole said, is by creating a false society reality. Cole refers to this as "the Instagram catastrophe."

"People are so motivated to misrepresent themselves and it's impossible to form any kind of honest communication that would actually provide a sense of being known and being well and being in community with one another," Cole said, adding that this is going to leave humans "emotionally and socially starving."

Clearly, Cole is very passionate about this. I can hear it in his voice, the frustration and the fear. His worry carries over to me as I listen to him say that he believes humans are facing "the most complicated traps we've ever set for ourselves as a species." As a health and science journalist, I'm aware of the flaws in some of the research suggesting that social media and technology are solely to blame for today's mental health problems. I reached out to Dr. James Doty, a clinical professor of neurosurgery at Stanford University and founder and director of the Center for Compassion and Altruism Research and Education (CCARE), and ask if, like Putnam theorized, the loneliness crisis is the result of our free time being privatized and individualized by technology. He said there was no question about that. He added that Facebook, Google, and the other technological entities don't care about their consumers. "Their only interest as a publicly traded company is to make money, right?" he remarked.

"They have a whole team of psychologists, neuroscientists, who figure out how to keep you on your device," Doty told me, adding that conflict is often part of the process to keep us on our apps. "It's highly manipulative, it makes people unhappy, and it's very disturbing."

In 2018, 50 psychologists signed a letter accusing psychologists of working at tech companies and using "hidden manipulation techniques" and sent it to the American Psychological Association, demanding the APA take an ethical stand.[12] While so many American adults and kids are spending time on their screens, there's a shift happening in how we discuss who's to blame. While parents often take the heat for not being able to control their child's screen time, which is a constant battle, what's often left out of these discussions is that these apps and devices are intentionally designed to keep people on them. In fact, there's a whole field of study dedicated to this called behavior design, or "persuasive technology." As Ramsay Brown, neuroscientist and cofounder of the artificial intelligence/machine learning company Boundless Mind, said in a *Time* article, "Your kid is not weak-willed because he can't get off his phone. . . . Your kid's brain is being engineered to get him to stay on his phone."[13]

And it's working. In 2022, the nonprofit research organization Common Sense Media found that overall screen use among teens and tweens increased by 17 percent between 2019 and 2021.[14] The survey found that, on average, 8- to 12-year-olds spend about five-and-a-half hours a day on screens, and 13- to 18-year-olds spend about eight-and-a-half hours a day on screens. Scientists like Cole say this is problematic. "We're the most intensely social species by far, this is what we are all about, this is why we have done as well as we have as a species," Cole emphasized to me. "We are literally decomposing with technology right now." He said he doesn't mean to "overdramatize" what's going on, but humanity is in a "moment of peril." "The loss of humanity is very much a real possibility," he warned.

· · ·

Scrolling, liking, commenting, thumbs up, thumbs down. I'm not sure when exactly it happened, but in the late 2010s it seemed like people in San Francisco who rode the public transportation system, known as Muni, did everything they could to avoid eye contact with each other. They swayed back and forth, from stop to stop, with their necks craned downward, staring at their phones, like crows on a cable line hunting for scraps. Here we all were hurdling underground, all headed in the same direction, and yet we could barely acknowledge each other. This disconnect, in a very connected world, has increasingly disturbed me.

· · ·

The nature of loneliness is nuanced. According to the late John Cacioppo, a researcher at the Center for Cognitive and Social Neuroscience at the University of Chicago who pioneered much of the relevant research we know about loneliness today, loneliness is a "subjective experience" that results from a discrepancy between one's actual social connections and one's desired social connections. It's a perceived gap between quality and quantity. Loneliness is the grief that's felt in wanting to feel more connected and being unable to obtain that connection.

In his book *Loneliness: Human Nature and the Need for Social Connection*, Cacioppo explained that everyone feels the "pangs of loneliness" throughout their life.[15] It can be as brief as being the last one chosen for a team on the playground or as severe as losing a spouse. "Feeling lonely at any particular moment simply means that you are human," Cacioppo wrote, arguing there's a genetic need for some people to have more connection than others, perhaps predisposing them to loneliness throughout life. Loneliness,

Cacioppo wrote, is a "social pain" that is meant to protect the individual "from the danger of remaining isolated." Knowing this is a normal part of the human experience, Cacioppo said that loneliness becomes an issue when "it settles in long enough to create a persistent, self-reinforcing loop of negative thoughts, sensations and behaviors."

Notably, many believe that the state of loneliness that's causing an epidemic today is fairly new. British historian Fay Bound Alberti argued in her book *A Biography of Loneliness: The History of an Emotion*, published in 2019, that modern loneliness didn't exist before the 19th century.[16] In fact, she calls today's loneliness a "modern emotion." In the preface, she explains that loneliness could have been seen as negative in previous centuries, but overall, those who chose to be alone and intentionally disconnected from society did so without feeling lonely because the spiritual and philosophical framework of the time was different. There is a difference between loneliness and solitude. As Cacioppo wrote, being physically alone does not necessarily mean being lonely. Over the course of human history, solitude has been highly regarded. Ancient humans sought out solitude as a mythical place of creativity and self-discovery. Religious hermits were individuals who chose solitude in wild places. Jesus was known to spend nights alone, praying.

Historians suspect that even people who appeared to relish modern loneliness, like Henry David Thoreau, suffered from it (despite Thoreau's celebration of solitude in *Walden*). In *The Journal of Henry David Thoreau, 1837-1861*, Thoreau seemed to express disappointment in his connections with friends.[17]

According to Cacioppo, there are three components that determine who suffers from modern loneliness. First, it's a person's level of vulnerability to social disconnection. Second, it's a person's ability to self-regulate. Finally, it's how a person mentally reasons and sets expectations about our social world. The final component is difficult because Cacioppo says that loneliness can create a negative feedback loop that is hard to break. "When loneliness takes hold, the ways we see ourselves and others, along with the kinds of responses we expect from others, are heavily influenced by both our feelings of unhappiness and threat and our impaired ability to self-regulate," he wrote.

Despite these nuances, researchers have found a successful way to measure loneliness in individuals. In 1978, researchers released a psychological assessment tool called the UCLA Loneliness Scale, which included a list of 20 questions meant to measure a person's perception of loneliness and social isolation.[18] Amid iterations and revisions, it is still a widely used measure today. After a revision in 2004, researchers boiled down the scale to three questions, to be used in instances where a measurement is needed more immediately. The questions are:

> How often do you feel that you lack companionship?
> How often do you feel left out?
> How often do you feel isolated from others?

Response options are "hardly ever" (1), "some of the time" (2), and "often" (3). Someone who scores a nine would be considered to be on the "most lonely" side of the scale, as opposed to someone who scores a three and is "least lonely."

In 2018, the health insurer Cigna used the 20-question UCLA Loneliness Scale to survey 20,000 adults online across the country.[19] Scores on the full UCLA scale range between 20 and 80, and if a person scores 43 they are considered lonely. According to the survey, the average loneliness score in America is 44, suggesting that "most Americans are considered lonely." The survey also found that the younger generation was lonelier than older people—and this was before the pandemic.

In October 2020, Harvard psychologist Richard Weissbourd and his colleagues found in a preliminary survey that 43 percent of young adults reported an increase in loneliness since the outbreak of COVID-19. In the survey, about half of the lonely young adults reported that no one in the past few weeks had "taken more than just a few minutes" to ask how they were doing in a way that made them feel like the person "genuinely cared."[20] It wasn't just a pandemic problem, as *Vice* declared in 2022: "Young People Are Lonelier than Ever."[21]

"We tend to think that young people are thriving and living wildly fun lives and the reality is they're not hooking up very much, a lot of them are lonely," Dr. Weissbourd told me over Zoom. Running his hand through his full head of silver hair, he explained how social media specifically has created a new kind of loneliness among young people, one that creates a gap in reality between quantity and quality of connections. "The constant comparing of yourself with people who are with friends and look joyous and look connected," he said. "I think these relative comparisons can really deepen loneliness."

Of course, it's not just young people. Loneliness researchers tell me they believe younger people *and* older people are the most affected. Even if middle-aged people score low on the UCLA Loneliness Scale, they could have higher scores in the future. That's because older adults in the U.S. are at a higher risk of loneliness and social isolation due to living alone, chronic illness, hearing loss, and the loss of family and friends.[22] In terms of the loneliness created by social media and technology, Cole said he believes older people are a bit better equipped to deal with the changing world. But younger people don't have the chance to build the infrastructure needed to have "nutritious life experiences," even how life was 5 to 10 years ago. Certainly, many think this will increase the likelihood of younger people being lonely later in life.

A decline in societal health due to loneliness runs parallel to a decline in our individual physical health. Loneliness is just as deadly as smoking. Researchers estimate that the risk of death from chronic loneliness is comparable to smoking up to 15 cigarettes a day, and that is even greater than the risk of death associated with physical inactivity and obesity.

In 2022, the American Heart Association found that people who are chronically lonely have a 29 percent increased risk for a heart attack and a 32 percent increased risk for stroke.[23] There is strong evidence linking social connection and physical health—particularly, that a lack of social connection can lead to heart disease, the number one killer of Americans. Loneliness in childhood, measured by social isolation, is linked to higher obesity, higher glucose levels, and high blood pressure later in life. The stakes are high from a public health perspective regarding the health effects of loneliness.

"Given the profound consequences of loneliness and isolation, we have an opportunity, and an obligation, to make the same investments in addressing social connection that we have made in addressing tobacco use, obesity, and the addiction crisis," Vivek Murthy wrote in his 85-page advisory, emphasizing the seriousness of the situation. Loneliness is also linked to various psychiatric disorders, like depression, alcohol abuse, and sleep problems.

Scientists who study the health effects of loneliness argue social connection is just as important as food for the human operating system. Biologically, physical pain is meant to signal that something is off in the human body. When a human hasn't eaten food for a couple of hours, the stomach, pancreas, small intestine, and brain all secrete a hormone called ghrelin. This activates receptors in a part of the brain to signal that nutrition is needed. It's how we know it's time to eat. If food isn't consumed in a timely manner, additional physical effects are felt, like shakiness and irritability. As Cacioppo explained in his book, loneliness is a warning that a human isn't feeling as connected to those around them as they need to be to feel safe.

However, when hungry, depending on their circumstances, people are usually able to help themselves. If they can't, people often come together to help feed each other, similar to how people come together after a natural disaster. If loneliness is our current crisis, why aren't we coming together in bounded solidarity to overcome it?

Cacioppo suggests it's because of the insidious nature of loneliness, which creates a catch-22. "Real relief from loneliness requires the cooperation of at least one other person, and yet the more chronic our loneliness becomes, the less equipped we may be to entice such cooperation," he wrote. If people don't receive the connection they are looking for, which requires a warm and cooperative response from another individual, it can start the negative feedback loop that Cacioppo identified. Cacioppo argued that not receiving the desired connection can trigger depression, hostility, despair, and impaired skills in social perception, which can lead to unhealthy attempts to mask the pain by seeking pleasure in whatever commodity, or person, is easily available. Additionally, offering support is hard to do for people who are lonely, which makes seeking connections difficult.

"But the real injury added to the insult is that, while this outward disruption is taking place, loneliness also disrupts the regulation of key cellular processes deep within the body," Cacioppo wrote. "This is how chronic loneliness not only contributes to further isolation, but predisposes us to premature aging." Cacioppo said it's a tragedy that loneliness not only makes us miserable but also physically sick and can lead to premature death.

The sly nature of loneliness can be observed in the brain. Recently, Elisa Baek, an assistant professor in the Department of Psychology at the University of Southern California, recruited college students who were starting their freshman year. Baek and her colleagues did magnetic resonance imaging (MRI) scans while the college students were watching various videos depicting everyday events they might encounter—for example, a sports game, a social party, or a scene from a movie—to see how their brains processed the events. The researchers proceeded to have each participant complete the UCLA Loneliness Scale. From there, they were able to see how lonely people were processing these clips, which were proxies for everyday life, differently. What the researchers found was that lonely people, on average, processed the videos dissimilarly, in their own unique ways. For people who didn't score high on the loneliness scale, their brains processed the videos in similar ways.

"It suggests that idiosyncrasies and how they respond, process, and view the world around them may be kind of why they feel this sense of not being understood by people around them," Baek told me in an interview. "They're literally processing the world in their own idiosyncratic ways."

Baek said that having a shared reality with others is really important to feeling socially connected. If lonely people aren't seeing their reality in the same way, being in the presence of other lonely people could actually be a risk factor for greater loneliness. "Being surrounded by people that don't view the world in similar ways as yourself may be a risk factor for loneliness," she said. "When your brain is responding in ways that are dissimilar from your group, you feel lonely."

A possible implication could be this: having lonely people sit in a support group and talk about how lonely they are might not be the way out of loneliness. Additionally, the solution might not be for a lonely person to go out and try to make new friends. "It suggests that having a lot of friends that you don't feel that close to can be a lot worse than having less friends that you don't feel as close to," Baek said. Instead, she said, the focus should be on how people can cultivate a shared meaning or understanding with each other. "Some ideas that we've been having were like, how can we cultivate a shared understanding in individuals that don't naturally see eye to eye on different things?" she said. "Like a novel path together or like a novel experience together."

4 The Coregulation Fix

I've been sound asleep for two hours, but a distant cry wakes me up. I peer over at the baby monitor that glows in the nighttime abyss like a lighthouse on the shore. It's only midnight, and my daughter is standing in her crib, crying for me, and babbling through her tears. It's almost as if she's awakened from a nightmare. Maybe she's hungry. Maybe she's scared. Maybe she's distressed that she fell asleep in my arms, and now I'm no longer in her peripheral vision. Maybe she's lonely.

Ten months into motherhood, my sleep debt could give the U.S. deficit a run for its money. As much as I want to turn over and go back to sleep, I physically can't. I feel a jolt of adrenaline inside my body. My heart beats faster. Everything feels more urgent. Before I can think more about it, I jump out of bed and head straight into her room. I swoop her tiny body over her crib and bring it close to mine.

The human body has 12 pairs of cranial nerves that stretch like tree branches from the bottom of our brain, down our spine, to various chambers.[1] Within this system, there is one very long nerve that helps control the body's autonomic nervous system.[2] Called the vagus nerve, its appendages touch the heart, lungs, and digestive organs. It's the autobahn, with tiny detours and side streets, between the brain and gut.

There are many sounds in this world that humans have become accustomed to hearing and as a result can easily ignore. But to a mother, a baby's cry is not one of them. That's by nature's design. Neuroscientists have found that a baby's cry activates primitive parts of the brain that are responsible for that anxious, fight-or-flight response that keeps me from being able to sleep through my daughter's cries in the middle of the night.[3] When this happens, the vagus nerve essentially "settles." How we respond, by relying on each other to put a brake on the fight-or-flight response, reactivates it and brings on a calming effect.

In 1975, psychologist Edward Tronick presented the "Still Face Experiment" to colleagues, a phenomenon where after three minutes of an "interaction" with an expressionless, still-faced mother, an infant grew concerned.[4] The infant then proceeded to make attempts to interact with the mother. When these attempts failed, the infant withdrew.

"But the baby's reaction is contingent on the intonation of the mother's voice," Stephen Porges, who is the founding director of the Traumatic Stress Research Consortium, explained to me over Zoom. If the mother's voice is calming, the baby's heart rate drops and the distressed behaviors get depressed. Porges said this is an example of how humans have roots in the "sociality of a mammal." "It's not what people say that we're attracted to—it's how they say it," he said.

Porges is interested in this because his primary focus of research is on the autonomic nervous system. In 1994, he proposed a theory known as the polyvagal theory, which is a framework to better understand the evolutionary role of the autonomic nervous system, the human response to stressors, and the cues of safety and threat.[5] Per the polyvagal theory, our autonomic nervous system operates in three hierarchical states. The first, in response to extreme threats, is going into a state of freeze. When this occurs, a person can be immobilized and completely unable to move, which can manifest as depression. The second is a state of fight or flight, which happens when there is a perceived threat. The final, and most desirable, state is a calm one where people feel open and safe, and are easily able to connect with one another.

Porges believes if we can create a positive feedback loop, one in which an activated vagus nerve is the default mode instead of it constantly being in a state of threat, hopeful change could happen in society. "If the autonomic nervous system is in its homeostatic state, it's an enabler for us to see the world in a more positive way and for sociality to be spontaneously emergent," he told me, elaborating that being social then can act as a "calmer." "When we are feeling safe enough, sociality comes out, and we calm others and that comes back on us and keeps us calmer."

This is done through a process called coregulation, and it's what happens when I calm my daughter. Only in that moment do I find relief, a sense of calm. My heart rate slows. I even start to feel sleepy again as I nurse her back to sleep. Together we have found our safe space. It's the circle of coregulation, or what I like to call the coregulation fix.

• •

Coming from a family of doctors, it made sense for Jonathan Leary to want to become one, too. As he prepared to take the Medical College Admission Test (MCAT), he had an epiphany: he didn't want to become a doctor. Not because he didn't want to help people, but because he feared the lifestyle of being a doctor. During his premed years, he saw how many doctors were struggling. They were sleep deprived, he said, and rarely got to spend time with their families. Despite working so many hours, they seemed to actually have very little one-on-one time with their patients. He made a personal decision to stop pursuing a career as a doctor and instead pursue one as a chiropractor. In the state of California, a chiropractor can be a primary care physician, with a few limitations, like prescribing medication.

On Sundays, he had time to work on a business plan. After writing 158 pages describing his vision, he went to the bank and asked for a loan. As he recalled the story to me, he laughed at the naivety of the situation. Unsurprisingly, the bank denied him a loan, so instead he opened up a concierge service as a sports medicine practice, focusing on surgery prevention and chronic pain rehabilitation. Within three months, he had two very high-profile clients. His practice took off. This led to traveling the world with top artists, royal families, and famous actors as their chiropractor. However, after a couple years of being a private chiropractor to the stars, one of Leary's clients asked him if he was doing everything he could to change health care, which was his goal when he first started his career. He dusted off his business plan, which eventually came to fruition as Remedy Place. Located in New York City and Los Angeles, the world's "first social wellness club" has become famous in part due to its celebrity clientele, including Kim Kardashian. It's not just another health club, but rather a place where social connections are also part of the treatment. Leary told me instead of meeting a friend for a drink, people are encouraged to take an ice bath at Remedy Place.

I first became intrigued by Remedy Place because of it being dubbed the "first social wellness club." The hype surrounding it confirmed what I had a hunch about all along: one-off yoga retreats, crystals, and pricey meditation classes that were marketed to middle-class millennial women as forms of "self-care" were no longer working, at least in a sustainable way. Late-stage self-care peaked right before the COVID-19 pandemic. Refinery29's 2019 blog *Feel Good Diaries*, which chronicled the physical and mental wellness routines of women, their costs, and whether or not these self-care rituals actually made them feel good, captured the height of it.[6] One posting described how a 28-year-old Los Angeles–based publicist, who made $75,000 a year, spent $2,003 on her wellness routine in a single week. This included a 50-pack of workout classes at Barry's Bootcamp, private surf lessons, lunch at Sweetgreen (a salad takeout joint), and falling asleep with the help of a pricey essential oil diffuser. Audrey Lorde's quote from her 1988 essay collection, *A Burst of Light*, splashed on Facebook and in workout studios: "Caring for myself is not self-indulgence, it is self-preservation, and that is an act of political warfare." While "self-care" had transformed into a consumption-based affair for people who had the resources to indulge in it, its history is rooted in Black feminism. The term was meant to portray self-care as a radical political act in a world that was hostile to you because of your identity.

Ashwagandha gummies. Consciousness festivals. Sound healing. Breath work. Tarot cards. Trying to open the heart chakra. Tracking daily steps. By 2015, millennials were spending twice as much as baby boomers on wellness.[7] For a long time, these consumption-focused activities took up a large portion of my budget in my attempt to feel good. But they also were misdirected, and disconnected from self-care's true value, and kept me less focused on engaging in and improving the community around me.

I'm not alone in thinking it's not a coincidence that America's self-help industry grew alongside the rise of the loneliness epidemic and technology over the past couple of decades. "Wellness, as we know it today, entered our culture around 2008, 2009, and it was the year the smartphone appeared to have hit a mass tipping point," Beth McGroarty, the vice president of research at the Global Wellness Institute, told me.[8] "The iPhone changed life so there was no line between work and life, we were suddenly bombarded by social media, the constant stress of notifications, and the comparison and loneliness because of social media."

In 1971, renowned positive psychologists Philip Brickman and Donald T. Campbell coined the term *the hedonic treadmill* to describe why consumption doesn't always lead to happiness.[9] It could be the reason why despite the growing wellness industry, the loneliness epidemic doesn't appear to be slowing.

Their theory was based on another psychological concept called the adaptation-level theory, which stated that an individual's judgment of a stimulus depended on their past experiences or encounters with it. The hedonic treadmill theory states that people have a baseline of happiness, and despite the influence of the goods they consume, their expectations will eventually rise and they will return to their previous state of happiness. Admittedly, it is a glass-half-empty kind of theory. But as the psychologists noted in their paper "Hedonic Relativism and Planning the Good Society," published in 1971 and which introduced the theory, there are pessimistic and optimistic ways to look at the pursuit of happiness through this lens. "The pessimistic theme is that the nature of AL [adaptation level] phenomena condemns men to live on a hedonic treadmill, to seek new levels of stimulation merely to maintain old levels of subjective pleasure, to never achieve any kind of permanent happiness or satisfaction," the psychologists wrote.

Years later, Brickman studied the hedonic treadmill theory in more detail. In their 1978 study "Lottery Winners and Accident Victims: Is Happiness Relative?," he and his colleagues measured levels of happiness among lottery winners and paraplegics.[10] People in both groups were asked to rate the amount of pleasure they got from everyday activities. The researchers found that the recent accident victims reported more happiness from activities like chatting with a friend, watching TV, and eating breakfast than the lottery winners did. As the researchers predicted, the findings suggested that the thrill of winning the lottery wore off. Since their expectations increased but their overall baseline of satisfaction remained the same, these simple pleasures would likely not have an impact on their happiness. "Thus, as lottery winners become accustomed to the additional pleasures made possible by their new wealth, these pleasures should be experienced as less intense and should no longer contribute very much to their general level of happiness," the researchers wrote.

According to a 2021 McKinsey report, the global wellness market is now worth more than $1.5 trillion.[11] However, it's becoming an increasingly crowded market, and wellness companies are having to get creative to stand out, like Leary's venture. As Leary worked with high-profile clients, the kind who could pay for any kind of private health treatment, he found they still weren't immune to the physical and mental health effects of loneliness. "When people are supersuccessful, their mental health does tend to struggle to a certain extent of success because human connection becomes less and less," Leary told me.

Or maybe, like Brickman and Campbell predicted, the novelty of new health experiences is deteriorating. Recently, some researchers have criticized the hedonic treadmill theory. But intuitively, there seems to be some truth to it. Through the wellness movement and technology, many people have been trying to pacify their loneliness through consuming wellness. But as shown by the hedonic treadmill theory, people eventually return to their baseline, which leads them to consume more.

When asked what's more helpful to his clientele, the treatments themselves or the emphasis on social connection, Leary said he believes it's both. Every treatment he offers, like ice baths or a sauna, he has tested on people and he says they work to promote positive health benefits. However, the environment, which is supported by human connection, also contributes to these positive health outcomes. He hypothesized that doing these experiences together will strengthen people's connections.

However, in terms of solutions to the hedonic treadmill, Brickman and Campbell said the only way to stop the hedonic treadmill is to get off it completely. "There are still wise and foolish ways to pursue happiness, both for societies and for individuals," they wrote.

• • •

Leary's theory that participating in wellness activities with other people as a way to improve one's health is supported by a growing body of research. In psychologist Catherine Haslam's theory called the social cure,[12] a sense of belonging and solidarity in a group can promote "adjustment, coping and well-being for individuals dealing with a range of illnesses, injuries, trauma, and stressors." In other words, it can build resilience. This is why at the end of Robert Putnam's book *Bowling Alone* he concludes, "As a rule of thumb if you belong to no groups but decide to join one you can cut your risk of dying over the next year in half." While joining a bowling league could be a start to ending loneliness, and improving individual and societal health, Haslam told me in an interview that a lot of her research has shown that belonging to one group can have positive health effects, but belonging to multiple groups is a way to sustain the effects. However, the solution to ending loneliness and improving both individual and societal health, to create durable solidarity, isn't to just simply join a bunch of social groups.

"It's not the case that just any group will do," Haslam told me. It has to be a group that provides a person with "a strong sense of identification of belonging and connection" and "life meaning, purpose and value."

"Somebody who has a lot of groups, but maybe they have a superficial influence on their lives, it isn't necessarily going to be health enhancing," Haslam said. "So it isn't about number, it's very much about quality, they're the ones that help you to make sense of your life, that give you a sense of meaning, and purpose."

One major issue, Haslam says, is that society isn't set up to support social connection, despite humans being a highly social species. Haslam added that, in her own research, personality doesn't predict who will benefit from social groups, but instead it's people who are experiencing social inequality and disadvantage. "Whether that's because of illness, disability, comorbid conditions, financial constraints, that robs you of opportunity," she says. "It's the opportunity to have the time to connect and the freedom to connect, as opposed to having to work three or four jobs." Haslam said there is another school of thought that suggests society has become too "insular" to connect.

"We need to invest in building those environments and places where people can connect," she said. "There are lots of communities and particularly disadvantaged communities where their safe spaces don't actually exist."

In one of Porges's papers, published in the journal *Infant and Child Development: Prenatal, Childhood, Adolescence, Emerging Adulthood*, he brings up a poignant point about human infants.[13] Unlike with reptiles that can glide, walk, and swim within hours after being born, for mammals birth is "not a transition into independence," he writes. Instead, it's an extension of a period of dependence. And as human infants mature, it does not lead to entire independence from others, either.

"Moreover, humans, as they become more independent of their caregivers, search for appropriate others (e.g., friends, partners, etc.) with whom they may form dyads capable of symbiotic regulation," he wrote. Indeed, our entire lives are centered around being together and seeking connections. In the same paper, Porges posited that the nervous system's purpose isn't to help us survive life-threatening situations, that is, through fight or flight, but to promote social interactions and social bonds in safe environments. In other words, it is meant to be in a state of calm and it thrives on our feeling safe with others.

But Porges told me in an interview that he suspects that in our modern world our nervous systems are always under threat. When I asked why, he said it's because we are constantly "under evaluation," which is "another way of saying our bodies are placed under threat." Fear, he added, is a massive motivator in many of our structured environments, like education, politics, and religion. Our nervous system works on a feedback loop, which isn't good when it feels under threat when we're feeling lonely. "An autonomic nervous system that is dysregulated in a sense is locked into a state of defense," Porges said. The only way for it to recover is for the person to feel safe—for society to feel safe. "If you're under threat, you're going to chronically retune your autonomic nervous system to be defensive," he told me. "And with that is a bias towards negativity and a bias for self-serving survival and a series of bodily function challenges." When the vagus nerve "settles," humans can be "an aggressive machine." Porges said that from the nervous system's perspective, humans are always in need of comfort, no matter what their age. "But our culture says deny that need, get over it, but you can't get over it," he said. "It doesn't matter how old you are, you want to be comforted, and it doesn't take a lot—the voice, the face, the script is very simple, but the issue is, can we implement it?"

It's more of that comfort, and what I call systemic caring, that we need to reverse the cycle.

. . .

"The bud / stands for all things, / even for those things that don't flower," poet Galway Kinnell wrote in his renowned poem "Saint Francis and the Sow."[14] Poetry critics speculate that the bud in Kinnell's poem represents the source of all creation. For some, it blooms. For others, it doesn't. In most cases, blooming depends on a person's season. No matter when the bud blooms in fullness, it's always there, inside us. This personification of the bud can be interpreted as an embodiment of the hope inherent within each individual, suggesting that the bud blooms only when we recognize the hope and goodness in others.

Similarly to how altruism acts as a catalyst for resilience after a disaster, the path out of loneliness is to be abundant with altruism and to prioritize it in our existing infrastructures. John Cacioppo knew this to be true, as he wrote in his book that altruism could be a pathway out of loneliness. "Altruism reinforced social connection, and social connection, along with the genetic dread of loneliness that is its flip side, helped our ancestors survive," he said. "Reaching out beyond one's own pain sounds like a tall order." But the road to success begins with small steps and manageable expectations, he said.

As a society, the U.S. is situated to face many crises, from future pandemics like COVID-19 to natural disasters like wildfires to heat waves exacerbated by climate change. While those crises need to be overcome on a disaster-by-disaster basis, if we don't make it through the loneliness epidemic, loneliness will keep us from coming together to face future threats to our health, as individuals and as a society. How we survive the loneliness epidemic will determine how well we overcome future crises. The first step out of loneliness is to prioritize kindness and caring for others.

II *The Alchemy of Altruism*

5 The Kindness Intervention

Lisa Luckett and her children woke up in their suburban New Jersey home.[1] The air was crisp. The sun brightened the sky, illuminating its blue in a strikingly noticeable way, the kind of way that makes people think or say, what a beautiful day! In many ways, it was like any other Tuesday, except Lisa's daughter wasn't feeling well. Luckett debated taking her to school. She did, but dropped her off a bit later than usual. When she returned home with her two younger kids, her phone rang. Her friend on the other end of the line hesitated. She asked Luckett which World Trade Center building her husband, Teddy, worked in. "The one with the antenna on it," Luckett said. "Why?" Her friend told her to turn on the news. A plane had just hit the building and demolished the top 15 floors. Her husband's office was two stories from the top, on the 105th floor.

Luckett's jaw dropped. Her heart sank. She held on to her two children as she watched the tragedy unfold in front of her in real time. The gaping hole in the tower. The smoke. The fire. Luckett's husband was one of the 2,996 people who died that day. The news initially threw Luckett into a state of shock, even though she said for years she feared her husband would die at work after the 1993 World Trade Center bombing.

Hours later, friends gathered at her house to comfort her as more details unfolded about the plane crashes. "I walked in my living room and I saw everyone so desperate and in pain and fear and all they wanted to do was help me," Luckett told me in an interview.[2] "But I couldn't get my head around it, the shock was crazy." While her grief could have immobilized her, she quickly found herself in what she described as a "place of love" by 11 in the morning after both towers had fallen. She described the feeling as an "incredible calm" that washed over her that could have been derived from a variety of sources. Perhaps the wave of calm was the shock in a different form. Maybe it was her having to stay grounded for her children. But she knew the only way to help those around her, which she felt a desire to do, was in fact to be vulnerable, to let them in and help her. When she did, she said she was overwhelmed by grace, gratitude, and humility. "These waves of love and positivity," she added. It's that place of love that helped her have the resilience to face a new normal that she'd have to endure for years to come.

Looking back, Luckett calls 9/11 a catalyst for her own healing, something that might not have happened if it weren't for all the kindness she received after that day. "When I hear 'Never Forget,' I never forget the beauty and the grace and the compassion and the resiliency and the incredible strength of the human spirit that came afterward," she said. Today, one of her only wishes is that people would treat everyone like how they treat 9/11 families. In fact, it led her to an epiphany years later. "We can solve all the world's problems if we just treat each other the way we wish we were treated," she said. "When we make random acts of kindness, we're actually filling our own souls."

The word *kind* is derived from the root word *kin*, which means "family." It's believed the word *kindness*, which is used to describe generous behavior, is in part a reminder that humans are all connected, like family, and need each other.

· · ·

For a long time in the field of psychology the primary focus of research was on studying negative emotions, such as sadness, anger, jealousy, and shame. In 1998, psychologist Barbara Fredrickson proposed the groundbreaking broaden-and-build theory of positive emotions,[3] arguing that positive emotions could contribute to a person's psychological health and mental well-being, and making the case that researchers needed to shift their area of focus from negative to positive emotions.

A couple of years later, as the psychology world still buzzed over Fredrickson's paradigm shift, psychologist Sonja Lyubomirsky said Fredrickson's theory opened up a new world of possibilities in terms of mental health interventions.[4] Why couldn't positive emotions be used to improve health, resilience, and overall mental well-being? Why was psychology just focused on reacting to and treating the so-called negative emotions? Since then, Lyubomirsky's career has largely focused on studying what she refers to as "happiness interventions," which include acts of kindness, inspired by the wealth of research in psychology that has shown kindness is good for mental, emotional, and physical health and building resilience to endure the heartaches of life.

In one of Lyubomirsky's landmark studies, participants were given a choice: perform five acts of kindness in one day or spread out five acts of kindness over the course of one week.[5] Lyubomirsky found those who performed all the acts in one day exhibited a greater increase in short-term well-being than those who spread them out. This finding, Lyubomirsky said, suggested that the timing of acts of kindness might be critical to their effect on well-being if they were to be medicalized in the future. She hypothesized that spreading out the acts of kindness may have diminished the novelty of the action, suggesting that variety might be key to a successful kindness intervention.

Through her studies, the evidence became increasingly convincing that there was a connection between performing acts of kindness and improving well-being. "Kindness can jump-start a whole cascade of positive social consequences," Lyubomirsky said. "Helping others leads people to like you, to appreciate you [and] to offer gratitude."

Building on her work, researchers in Canada tried to see if engaging in acts of kindness could improve social connections for people who struggled with social anxiety.[6] For their experiment, the researchers recruited undergraduates with elevated symptoms of social anxiety. Participants were randomly assigned to one of three groups. The first group of participants were assigned to record life events. The second group of participants engaged in behavioral experiments, like identifying their safety behaviors in social situations, such as avoiding eye contact. Researchers asked the third group to perform three acts of kindness two days a week. Once again, the researchers found that the group who had performed acts of kindness reported the most improved well-being and were better able to overcome their social anxiety as compared to the control groups.

As a clinical psychologist, David Cregg spent a fair amount of time in graduate school ruminating over the aspects of life that can sustain a fulfilling one. He did this by poring over existing research in the field of positive psychology, like the studies I mentioned above. "What I kept finding over and over and over again was that it's really the social connection with other people that seems to be the most potent ingredient for predicting flourishing in life," Cregg told me in an interview. At the same time, he wondered, what are we doing to address loneliness, to address skyrocketing rates of mental health problems? Sure, cognitive behavioral therapy (CBT), an emerging and popular type of psychotherapy that helps people change their negative thought patterns, has demonstrated success at reducing depression and anxiety symptoms. In part, that's because of the nature of CBT. It's a type of therapy that is typically conducted between a patient and a therapist. CBT therapists often prescribe homework activities, like worksheets, that are discussed in each session. Psychologists believe CBT is effective because it increases a person's awareness around their negative thought patterns. The therapy helps people see challenging situations more clearly. It gives them the hope and agency they need to see that negative cycles can be changed. Compared to doing nothing, CBT has been shown to improve social connection, but its effects haven't been as significant on anxiety and depression. As Cregg couldn't shake, research shows there is a strong link between health, a life well-lived, and social connections.

Lyubomirsky's research, and the research from others in the field of acts of kindness, inspired Cregg to do his own study to see how kindness, CBT, and social activities stacked up against each other. Cregg and his colleagues recruited 122 people who had moderate to severe symptoms of depression, anxiety, and stress.[7] The participants were split into three groups. The first group was assigned to use a CBT technique called cognitive reappraisal, where people keep thought records at least two days a week in an effort to disrupt negative thought cycles. Researchers instructed the second group to plan social activities for two days a week, like a dinner or outing with friends. For the third group, researchers had them perform three acts of kindness two days a week for a month. The researchers said the acts of kindness could be as "big or small" as the participants liked. They just had to "benefit others or make others happy, typically at some cost to yourself in terms of time or resources."

Prior to executing the experiment, Cregg briefed those who were assigned acts of kindness on how they could benefit the most from the assignment. For example, he said that studies like Lyubomirsky's suggest variety in acts of kindness matter. In other words, don't do the same act of kindness for the entire experiment or it might feel more like a chore. Regarding the frequency of acts of kindness, Cregg said he believed, based on previous research, that three acts of kindness two days a week for four weeks, which was the duration of the Canadian experiment, was the "sweet spot." "That way you're getting that more consistent reinforcement throughout the week," Cregg said. "But if you're going over that it can start to overwhelm people."

Cregg told me that he and his colleagues entered the study expecting all three interventions to benefit the participants' mental health, and they were right. The CBT intervention, prescribing social activities, and acts of kindness all showed measurable improvements in depression, anxiety, stress symptoms, and life satisfaction. However, the acts of kindness intervention improved depression, anxiety, stress symptoms, and life satisfaction more than the other two interventions. Like Cregg said, it's not that the other two didn't work, but they weren't as effective as acts of kindness. And there was a bonus: the participants who performed acts of kindness showed more measurable improvements in feeling socially connected, too.

What is it about acts of kindness that makes them prevail over CBT or just going out and socializing with a friend? When asked, Cregg pointed to the self-determination theory in psychology, which suggests that humans have three basic psychological needs.[8] The first need is autonomy, meaning that a person feels as if they are freely choosing their actions. The second is competence; a person wants to feel good about what they do or have the sense that their actions are making a difference. The third is relatedness, which is essentially social connectedness. Cregg said he believes that acts of kindness touch on all three. A lack of getting these needs met can negatively affect a human's well-being. When a person gives to another person, they're choosing to do something. It's not because they're being paid to do it, like the job at work that they have to do. When a person performs an act of kindness, they are able to see the impact and feel like they are competent at something. Additionally, it acts as a way for people to connect with each other in a unique way. "I think there's something more intimate about me going out of my way to do something special for you, where I don't expect anything in return," he told me.

Cregg told me he was excited about his results because acts of kindness are easy to do. They don't have barriers, like requiring money or a lot of time. It could be an easier intervention than asking people to keep CBT-like thought records, he added. "As people do kind actions, it sort of takes the focus off themselves and places it onto other people," he said. "It's sort of like I'm getting absorbed and helping others and I'm less fixated on my own problems." It's the same shift in mindset that Luckett experienced after losing her husband on 9/11, the one that helped her build resilience to persevere.

Five weeks after Cregg's study, he reached out to participants to see who was still performing acts of kindness as a way to improve their stress, anxiety, or depressive symptoms. Even if they weren't experiencing improved symptoms, were people performing acts of kindness as a way to build resilience? As a form of self-care? Seventy-five percent reported that they were. Cregg said he didn't know if they were doing three acts of kindness twice a week, though. "But what we found is that the improvements that they had during the active portion of the study remained," he said. "So that would suggest that even maybe more casually doing these acts of kindness after a more intensive period of it, making a habit of it, seems to still help with our mental well-being."

Throughout Amit Kumar's career as a psychologist, he has conducted research to explore the relationship between personal well-being and money. In one of his studies, he asked one group of participants to think about an experience they had just purchased, like a vacation or concert.[9] He asked the second group of participants to recall a material possession they had just purchased, like clothing or a car. At that point, all participants were asked to partake in a game where they were given a small amount of money. Kumar and his colleagues told the participants in both groups they could decide how to spend that money: they could either allocate it among themselves, keep it for themselves, or give it to an anonymous person who wouldn't know where it came from. What Kumar and his colleagues found was that the group of participants who had just recalled an experience they had purchased, instead of a material possession, were more likely to give the money to the anonymous stranger.

What Kumar discovered is that material possessions don't make people happy—experiences do. And it's not just happiness and fulfillment that stems from experiences, but also altruistic behavior.

In an interview, Kumar told me this finding from his study suggested that there are some types of consumption that could promote "other-oriented behavior." Time and time again in his own research, Kumar has found that kindness and generosity make the performers of the acts of kindness "substantially happier," not material possessions. "That's something that we measure in all of our studies, and that shows up as being a robust result in all of our studies," Kumar said.

In another one of Kumar's experiments, he and his colleagues recruited 84 participants in Chicago's Maggie Daley Park. Study participants were given the choice of giving away a cup of hot chocolate to a stranger or keeping it for themselves. Admittedly, it would have been a tough choice for me in the face of Chicago's brutal winter wind. Still, 75 of the study's participants agreed to give the cup of hot chocolate to a stranger. Notably, researchers reported mood boosts for both givers and receivers.

One unique part of Kumar's research is that he often measures both the satisfaction of the giver and the receiver during acts of kindness, not just one or the other. Interestingly, Kumar found that the people who gave away their hot chocolate underestimated how happy they made the strangers feel by giving away their drinks. They said they expected the recipients' mood to be an average of 2.7 on a scale of 1 to 5, when recipients reported an average of 3.5. "Our results suggest that we often think that our prosocial actions are going to be relatively inconsequential, even though they're not," he told me.

Kumar said this meant there are psychological barriers that might keep people from engaging in acts of kindness in their everyday lives. It's not that people don't want to be kind. In fact, he brought my attention to a piece of data people often miss from his series of experiments. "We found that people told us they wanted to perform *more* random acts of kindness in their lives," he said.

In their book *On Kindness*, psychoanalyst Adam Phillips and historian Barbara Taylor argue that we live in a culture that deprives us of relishing the pleasure of kindness.[10] "There is nothing we feel more consistently deprived of than kindness," the authors write. And yet, "the unkindness of others has become our contemporary complaint." Instead, living a kind life has become somewhat secretive, in part because kindness equals vulnerability, and vulnerability in our culture is seen as weakness. But by living like we are, we are only hurting ourselves. "In giving up on kindness—and especially our own acts of kindness—we deprive ourselves of a pleasure that is fundamental to our sense of well-being," Phillips and Taylor write. A kind life is the one we would be more inclined to live if only we could get out of our own way.

Joan Morgenstern built a robust career as an educator. As the director of a childcare center, she was required to provide guidance to parents. During her tenure, she noticed a trend. Despite getting solicited advice, parents frequently responded by thanking her and then stating that they were going to consult with their pediatrician. "And what it said to me is that the voice of the medical providers is different from the voice of educators," Morgenstern told me in an interview. As fate would have it, a position opened up at a local pediatric practice that was perfectly suited for Morgenstern: Director of Parent Education and Community Engagement. In this role, she would work one-on-one with parents in private coaching sessions to support them on their parenting journey. Morgenstern also knew about the Greater Good Science Center, an institution that focuses on the study of the psychology, sociology, and neuroscience of well-being at the University of California, Berkeley. When she learned that they were offering grants to organizations that were interested in promoting prosocial behaviors to parents and their children, she decided to apply, with a focus on prescribing acts of kindness to children to improve their well-being. "We want kids to read, and we know better how to help them read, but what do we do about helping them be kind?" she told me. "How can we be intentional about helping kids develop that empathy muscle?"

When Morgenstern started her position at Senders Pediatrics, she applied for and received the grant. Dr. Shelly Senders, the founding pediatrician of the practice, did not need to be convinced of Morgenstern's plan, in part because he had done a similar, alternative health mission two decades before. In an effort to improve literacy rates for children in his community, he started a program called A Daily Dose of Reading. In this program, he and his colleagues handed out prescriptions citing books for children to read. Could the same be done with kindness?

One of Morgenstern's programs that emerged from the grant was the Be KIND (an acronym for "kin-initiated nice deeds") initiative, where the practice handed out $100 micro-grants for kids to generate an idea on how to be kind. Morgenstern said that the goal isn't to get "grandiose" ideas, but rather to make children feel empowered to see that they can make a positive difference in a small way. Morgenstern said children crave feeling a sense of autonomy, a sense of having power—hence, the stereotypical toddler tantrums. While it's hard to give children a sense of power in this world, giving them the opportunity to perform an act of kindness could satisfy their need for it.

In addition to the Be KIND micro-grants, Morgenstern started a Family Kindness Festival, where the mission is to create an annual event to encourage kids to participate in acts of service in their communities. At the festival, there are two ways for kids to connect with kindness. The first is that student activists, usually from the Be KIND micro-grant initiative, have an opportunity to find other kids who will champion their cause. Another way is for kids to simply share their ideas on how to help others.

Dr. Senders shared a heartwarming anecdote from the Kindness Festival. One attendee named Derrick Smith said that he noticed none of his friends were reading because they didn't have any books. This gave him the idea to set up a station on the street where he and his friends would give out free books to kids and talk about the importance of reading. His organization is now called Boys Do Read.[11] Another child at a Kindness Festival said that she noticed hospitals had a lack of crayons. She started a nonprofit called Color Me a Rainbow and has since distributed 54,000 boxes of crayons to pediatric hospitals across the country. Morgenstern said her goal is to normalize kindness festivals, as a place where ideas like these can be generated and sustained from a young age. Film festivals and book festivals are common across the world, she said. Why aren't kindness festivals?

Morgenstern said she wants to shift the paradigm on kindness from being an "aside." Her latest project to do this is through what she calls "kindness boxes" that will encourage kids at Senders

Pediatrics to make a habit out of kindness. "I think maybe this notion that we can help children develop a habit around kindness could become kind of contagious in a good kind of way," Morgenstern said. "By defining it, very specifically, what habituated behaviors could get repeated over and over and become incorporated into a lifestyle." While she is still developing prototypes for the boxes, they currently include very short, actionable activities. For example, one is to "color a smile" for a nonprofit called Color a Smile. In this activity, a child colors in an image of a smile. The parent can mail the colorful smile to the nonprofit, who then sends it to people who could use a smile. Letters on their website from recipients will make even the heartless shed a tear. "Since I live alone, a colored picture from someone always makes me smile," one recipient said.[12]

• • •

For Dan Morse, a 10-year-old growing up in a suburb of New Jersey, 9/11 hit quite close to home. It became even more personal for him when, the next day, his dad took two weeks off work to go to ground zero and help people. "He literally went in while smoke was still rising in the background and approached police officers and firefighters and offered psychological counseling services," he told me in an interview. "I remember him coming home on the train with this soot-filled NYPD hat that was given to him, and really thinking my dad's a hero." It wasn't until years later that Morse realized his dad wasn't the only one who felt called to put a pause on their own life to help people at ground zero.

"So many people stepped up to volunteer to give their time to support," he said. "I found it so fascinating how when we're in times of crisis, humanity rises up to support each other, yet, when we're in these times of silent crisis, where there's millions of people depressed, lonely, and anxious, we don't always tap into that latent altruism that has power to transform our culture."

Throughout Morse's career he's been eager to find ways to do just that. Previously, he led an urban agriculture initiative to help young people grow their own gardens. He even started a restaurant where people could buy 10 meals and then receive a pass for a free local dance or yoga class. Throughout his various careers, he's spent a lot of time thinking about how to create an infrastructure to support social connection nationwide, which is how he came across the social prescribing movement in the United Kingdom. "They scaled their efforts nationally, millions of people were getting volunteer prescriptions, among other prescriptions, and I was so inspired that I wanted to start something in the United States," he told me. Today, Morse is the founder of Social Prescribing USA, whose mission is to "make social prescribing available to every American by 2035."

According to England's National Health Service (NHS), social prescribing is "an approach that connects people to activities, groups, and services in their community to meet the practical, social and emotional needs that affect their health and wellbeing."[13] It is essentially shifting from the idea of treatment being a medication or something that's delivered through a tube to an approach where doctors prescribe volunteering, kindness, art, or social connection. In England, health services are recommended to connect a patient to what they call a "link worker." In this meeting, the patient and link worker discuss the patient's health issues and what matters to them in life. From there, the link worker provides the patient with a social prescription after doing a holistic assessment, which can vary widely. In one case study from the NHS, a woman named Arabella Tresilian suffered from chronic fatigue and post-traumatic stress disorder.[14] Tresilian had maintained a successful career as a management consultant, and her general practitioner (GP) recognized that she needed more than medication to improve her health and well-being. Tresilian's GP referred her to a link worker, who did a holistic assessment to help identify the things that mattered to her and to see what options were available in the wider community. The link worker also looked at the systemic issues affecting her, such as finances and employment. Together, they decided that Tresilian would benefit from joining a choir. Three years later, she's still singing with the group, which has helped her obtain a more balanced life and be more involved in her community. "I work part-time, I'm involved in many activities and I feel good about the future," Tresilian said in the case study. "Social prescribing made all the difference to me and now I'm able to give back to my community."

Appropriately led by volunteers, Morse's organization is trying to build a similar movement to build awareness around social prescribing and make it accessible to medical professionals, which is how I connected with Senders Pediatrics.

In bringing the movement to the United States, Morse asked, "What if instead of just prescribing pills and procedures, we prescribe people and purpose?" Morse pointed to the strangeness of drug commercials on television. They often involve someone having chest pain or some other health issue. A drug is promoted as the solution, and as the person healthily goes about their day, presumably after taking the drug, a voiceover starts to speak very fast and says that the drug being promoted might cause a slew of side effects.

Drugs are incredible, Morse told me, but so is social prescribing. He said someone might be prescribed to volunteer to help clean up trash, which could inadvertently lead to a "Trojan horse solution" because this also leads to more physical activity.

"Imagine 10 more million people volunteering in the United States," he said. "What would happen?"

As Morgenstern told me, though, there aren't many pediatric or medical practices prescribing kindness yet. In fact, it was Morgenstern's unique idea that landed her the grant to fund her initiatives in the first place. "When you say you can't find medical practices that do this, we know that because there was no competition," Dr. Senders told me. Notably, there is also a gap between the few medical practitioners implementing social prescribing and those studying it. As Morse works on igniting the social prescribing movement in the U.S., what can be done to prioritize acts of kindness in our culture today?

Perhaps it might take a little creativity, and once again, be a lesson from tragedy.

• •

On a Monday morning, clouds hovered over downtown Los Angeles. A rare few sprinkles of rain fell from the sky, providing relief before the day's unrelenting heat set in. After walking through airport-like security in a parking lot, I entered a pop-up tent on top of a parking garage with nearly 400 people. As we dispersed to our respective tables, I brushed shoulders with mascots representing the city's sports teams, and even some celebrities. Music pulsed through the speakers as everyone found their respective spots. Sure, fashion week was happening on the other side of the country. But this group of Angelenos had come together for a different kind of celebration. The mission of the day was to pack hundreds of thousands of meals for people facing food insecurity to commemorate the 22nd anniversary of 9/11. Alongside 18 cities nationwide, totaling nearly 30 million Americans, the volunteers I witnessed were gearing up to participate in a uniquely connected day of service in America. It's a day that has been officially recognized under federal law.

Historically, we remember national tragedies in well-intentioned, albeit predictable, ways. A moment of silence to remember the lives lost. A wreath placed on a grave. Stories shared from loved ones. On the anniversaries of national tragedies, we are encouraged to grieve and to remember. For a nation and culture that doesn't always take the time and space to heal (cue the countless "thoughts and prayers"), it's important to honor loss in these funereal ways. But what if it doesn't always have to be so somber? What if there was another way to carry on the memory of such an event, one in which participants could simultaneously heal and transform their grief? A way that would not only honor lives lost, but also benefit participants' communities at large? A day to remember the kindness, like Luckett said, and also perpetuate it.

In a way, that's what Jay Winuk, whose younger brother was killed while working as a volunteer firefighter at the World Trade Center, wanted for future 9/11 anniversaries.[15] He and his friend David Paine wanted good to come from a day that would be remembered for much loss. They wanted to turn future anniversaries into a day of volunteering.

The mechanics of the mission were well-run. First, a person poured dry lentils through a tube into a bag. Next, another person sealed the bag. Then, a third person (including me) placed a sticker on the bag. Another person packed each bag in a box. Within 30 minutes, we finished three boxes, totaling over 100 meals that had the potential to feed dozens of families for weeks. Not only did I feel good to do good, as cheesy as it sounds, but at that moment I felt safe in the presence of others working toward a goal to help others in the community.

U2's song "It's a Beautiful Day" played in the background as I asked a volunteer named Crystal Lawrence what brought her to the event. As an AmeriCorps fellow, she said she's a regular at events like these. But she also told me she believes that meal packing days are empowering to the community. They help remind us that we can be the solutions to the problems we see around us, the same ones that cause us to feel despair. "The 'for us, by us,' standpoint," she emphasized, adding these events are also opportunities for people to grow individually and communally. "And you really leave happier." Another volunteer recalled to me that after last year's event she felt "high." She even experienced sadness the day after as she missed packing in a big auditorium, helping others and making new friends.

When I asked another woman why she was volunteering that day, she told me it was an opportunity that her job at Fox News Corporation provided to her. As a lifelong progressive Democrat, in this divisive political climate, I typically wouldn't find myself mingling with someone who worked for Fox News. But for the day, as we placed stickers on bags of beans, we became friends. We were working together to help fellow humans. By the end of the day, we had packed nearly six million meals with our fellow meal packers across the country.

Later in the day, I asked Josh Fryday, California's chief service officer (whom you will meet again in a later chapter), what he thought about commemorating 9/11 through acts of service. He said he thought it was a way to briefly relive the sense of togetherness, the bounded solidarity, that people experienced on that September day in 2001. "What service and doing good shows us is that when we're united, when we help each other, when we serve each other, when we are connected to each other, we can get through anything," he said. "Today reminds us of that spirit, and it rejuvenates that spirit."

But then he paused, reminiscing that there was a sense of helplessness, too, after 9/11. As many may recall, at the time, President Bush's primary declaration to Americans in the aftermath of the attacks was to "go shopping." Fryday said he and others he's spoken with wished there had been an infrastructure in place at the time for people to be called on to help people. I asked him, does today help build that infrastructure for future crises? Yes, he answered. Civic engagement, he added, is like a muscle. "At a basic level we are connecting people here," he said. "Civic engagement is not something that we just want to turn off and on, it's a continuous muscle that we've got to keep strong, we've got to keep nourished, or it will go cold on us."

6   Better Together

On a tiny island called Cayo Santiago off the coast of Puerto Rico exists a colony of about 1,800 rhesus macaques. Each weighing about 20 pounds and known for their sand-colored fluffy tails, the monkeys that inhabit this island today are descendants of those brought over by primatologist Clarence Carpenter in the late 1930s. Since then, they have helped primatologists, evolutionary biologists, and scientists of all kinds better understand primate behavior in a unique natural laboratory setting. Neuroscientist Michael Platt is one of those lucky scientists who has been able to study them for over a decade, particularly with a focus on how their social environment affects their brains, how they make decisions, and the genetic underpinnings of their social behavior. When news broke in the fall of 2017 that Hurricane Maria, a Category 4 storm, was bound to make landfall, Platt and his colleagues were terrified.[1] They worried about what this would mean for their research and the monkeys who had given so much to science.

On September 20, 2017, the hurricane hit at a ferocious speed, pummeling the island with 170-mile-per-hour winds and torrential rains.[2] Platt and his colleagues waited several nail-biting days to hear about the assessed damage and potential mortalities of the monkeys. Upon their colleagues' surveying the scene by helicopter, a heroic effort at the time, they learned that two-thirds of the island's green vegetation had been wiped away. The freshwater cisterns that the monkeys relied on as a water source were destroyed. Through a collective effort, researchers were able to get back up and running fairly quickly, which positioned them to be in a unique opportunity: to see how the rhesus macaques would respond in the wake of a natural disaster. Specifically, the researchers were curious to see if the monkeys' social ties had shifted and if their behavior would turn more tolerant or aggressive. Considering the lack of resources and devastation, would the monkeys fight over strained resources in the quest to survive?

Since the researchers had over a decade of their social behavior documented, they'd be able to compare the monkeys' behavior from before the hurricane to that after the hurricane. For example, they knew that while these monkeys are highly social, they can also have very competitive streaks.

Previously, the researchers had relied on a study method that required researchers to follow each individual monkey for 10 minutes and report every action and interaction to study their behavior. Since the devastation was too big to support this kind of approach, researchers turned to another sampling technique known as the "scan method."

In this technique, an observer looks up every 30 seconds to record the interactions of every monkey around.[3] After adjusting for potential biases, like louder monkeys trying to grab the attention of the researchers, an analysis of their data showed that the monkeys' behavior had indeed changed after the hurricane. But instead of for the worse, it was for the better.

For instance, the monkeys appeared to be more tolerant of each other compared to the previous times. While the researchers expected the monkeys to rely on those they already had invested relationships with to cope with the ecological devastation, they found that the monkeys appeared to actually seek out new relationships and expand their social networks. A close relationship still had a lot to provide, but it was almost as if the monkeys experienced a realization that a social network where everyone is friendly enough is better for their overall survival than a network with just a few close friends. "What was amazing was that these monkeys immediately began to reach out and make more friends," Platt told me in an interview. "And everybody got connected with everybody in a dense web of interconnection." Fascinatingly, even monkeys who were previously characterized as socially isolated broke out of their lonely shells and made more social connections in the hurricane's aftermath.

While most of the monkeys survived the initial impact of the storm, the population experienced an uptick in mortality a month later. As time went on, researchers found that the monkeys who had more friends were more likely to survive in the damaged ecosystem for the following two years. And it wasn't just their physical habitat that had experienced rapid deterioration. Platt and his colleagues made another observation of the monkeys. Some appeared to have aged about two years. Monkeys in their teenage years were developing arthritis.

Five years later, the stronger and more tolerant connections among the monkeys were still living on. "The monkeys are still way less aggressive, way more tolerant, and more connected with each other," Platt told me. Indeed, it appeared that the monkeys experienced bounded solidarity and were able to transform it into durable solidarity. Why did it work for the monkeys, and why doesn't it for humans? That's one of a few million-dollar questions, Platt said. Other open-ended questions are these: Why did some monkeys appear to be able to overcome the difficulties of the hurricane more than others? Why did some show early signs of aging from the stress, and others didn't? In other words, why were some more resilient?

Social support is thought to be an adaptive response to extreme stressors, Platt said. This means that having strong social support before a tragedy can help organisms better resist stress damage. The implications for the monkeys could be this: those who had stronger social connections before the hurricane were able to cope better with the aftermath of the hurricane. Platt said that there's a lot of compelling research that shows more social connections can act as a buffer in the brain against stress responses. It can help people get through tragedy, disaster, and trauma. It can keep people's brains young, in a sense. "And if you have a younger brain, you're probably going to be able to navigate life better too, so it's a feedback loop," he said. "When your brain is older, you're not going to be able to navigate a lot of those complexities."

We know from research on monkeys and humans that having more social support enables resilience, Platt said. But the big open question is, how?

• • •

Perhaps the first step to understanding how having more social support enables resilience that can be observed in the brain, it's best to understand how stress affects the brain. To find an answer, I reached out to cognitive neuroscientist Dr. Julie Fratantoni, who is also one of the leaders behind the BrainHealth Project, a 10-year longitudinal research study seeking to define, measure, and improve brain health.

She said broadly speaking, chronic stress kills brain cells in the hippocampus, which is the part of the brain responsible for learning and memory. "Your neurons literally die," she told me. When that happens, it can become more difficult for people to learn and remember things. Stress can also affect the brain's frontal networks, which are responsible for executive functions like planning, judgment, organization, and problem solving. Higher-order thinking, she said, makes humans different from other animals. When stress shuts down this part of the brain, humans are then forced into survival mode. Further, this shutdown narrows down our options to regulate ourselves. It turns the human brain into a reptilian one, like Stephen Porges described in chapter 4, and activates the sympathetic nervous system, putting us into fight-or-flight mode—the same one we can get stuck in when we're chronically lonely. Dr. Fratantoni said one way to turn the prefrontal cortex back on to a less stressed mode, one that can think more clearly, is through curiosity. When I asked if altruism could be a way, she said it's possible because there are a lot of similarities between kindness and curiosity. Both, she said, are an "open posture." While kindness is hard to access in the immediate aftermath of stress, just as it can be when someone is chronically lonely, it could be a shortcut to bringing the prefrontal cortex back online.

What do we know about what happens in the brain during an act of altruism? In 2006, neuroscientist Jorge Moll and colleagues provided some of the first evidence to demonstrate what happens in the human brain when a person gives selflessly to another person. In their experiment, the researchers scanned participants' brains using a functional MRI as participants made decisions about whether to donate money to a charity, oppose donating to a charity, or receive the monetary reward themselves.[4] As they scanned the brains of participants while making decisions, researchers found that those who chose to keep the monetary reward for themselves experienced activity in the mesolimbic reward system, including the ventral tegmental area and the ventral striatum.

The mesolimbic reward system, sometimes referred to as the reward pathway or the mesolimbic pathway, is responsible for releasing dopamine, a neurotransmitter that allows us to feel pleasure and satisfaction. It also plays a role in motivating us to want more, like food and sex. This reward pathway regulates motivation, reinforces learning, and activates incentive salience, which is a cognitive process that makes us experience "desire" or "want." Its job is to motivate us to repeat behaviors that are needed to survive. Notably, this reward pathway also plays a significant role in the neurobiology of addiction. The findings in Moll's study did not come as a surprise. Of course, receiving the monetary award felt good and activated the desire to want more. However, when scanning the brains of those who gave the money to charity, scientists saw that these people experienced even more activity in this reward pathway. This finding suggested that giving to other people could provide more pleasure—per the brain's reward system—than doing something that feels good for oneself. Notably, donating the money also activated the subgenual area of the brain, a circuit of the brain that scientists know is rich in serotonin and plays a role in social bonding, which was not activated when the study's participants chose to keep the money for themselves.

Dr. Richard Davidson started his career, like many in his field, with a focus on the neuroscience of depression, fear, and anxiety.

Similarly to what Platt wondered about the monkeys, Dr. Davidson was curious to know why some people are more vulnerable to life's challenges and stress. And why are others more resilient?

An encounter with the Dalai Lama in 1992 changed the trajectory of Dr. Davidson's research. When they met in Madison, Wisconsin, the Dalai Lama challenged him to focus his research on something more "positive." He asked him this question: why can't you use the same tools in neuroscience that you use to study fear and depression to study kindness, compassion, and altruism? Dr. Davidson took this challenge seriously. Over the past three decades, he's published dozens of papers on kindness, compassion, and altruism and has even studied the brains of Buddhist monks. "One of the things that seems to be true is that when we do things for ourselves those experiences of positive emotions are more fleeting, and they are more dependent on external circumstances," Dr. Davidson told me in an interview in 2017.[5] "When we engage in acts of generosity, those experiences of positive emotion may be more enduring and outlast the specific episode in which we are engaged."

Thanks to research like Moll's, scientists know that the region of the brain that is impacted the most during an act of kindness is the ventral striatum, which is important for motivation and positive emotions. It's the same region of the brain that is occupied by dopamine, which we know as the "happy hormone," but Dr. Davidson cautioned me there are more molecules at play here. "There are hundreds of molecules involved and it's very inaccurate to try to pin this on one or two chemicals, the change is likely to be far more complicated," Dr. Davidson says, adding to the mystery of how helping others makes us feel good and builds resilience.

When it comes to the human brain, today's scientists are just understanding the tip of the iceberg, so to speak. There are billions of neurons in the human brain and the reality is that scientists have little understanding of what they do when we move, when we love, when we dream, when we give, and when we sleep. Brain scans tell a small part of the story. They provide clues and hints. Still, the human brain remains somewhat of a mystery that needs to be solved. Part of the difficulty in solving it is a lack of technology, and the other part is a lack of human brains to study. That being said, there have been significant advances in understanding that brains can change, meaning that the brain is actually very flexible. This quality is known as neuroplasticity, and neuroscientists have seen time and time again that the brain has the ability to change and adapt in response to experiences.

Building off the idea of neuroplasticity, Dr. Davidson conducted a study in 2013 to see if humans could be trained to be more compassionate, the emotion of caring for someone who is suffering, which many scientists say motivates altruistic behavior.[6] Can we become more compassionate, and in turn exhibit more habit-based altruistic behavior, if we train our brain to be more compassionate? The researchers assigned two groups of participants to activities that they were told would improve their overall well-being. The first activity was to cultivate kindness, compassion, and generosity toward others by participating in a compassion meditation. In the meditation, participants envisioned a time when someone was suffering and then wished that their suffering was relieved. They did this with three different categories of people. The first was a loved one, someone they could easily feel compassion for, like a family member or friend. Then they practiced compassion for themselves and a stranger. Finally, they practiced compassion for a "difficult person," someone they had a history of conflict with, like a competitive coworker or headstrong family member. The second group of participants engaged in cognitive reappraisal instead of compassion meditation. In this activity, people worked on reframing their negative thoughts. Both groups took part in their activities for 30 minutes per day over the course of two weeks. Dr. Davidson and his colleagues were specifically curious to see if emotional states could be changed in a short amount of time—if, like physical exercise, compassion was a muscle that needed to be fine-tuned every day.

To see whether or not the compassion meditation worked, researchers needed to learn if those who did the compassion meditation were more altruistic at the end of the experiment than those who did the cognitive reappraisal activity. To find the answer, researchers asked the participants to play the Redistribution Game on the internet with two anonymous strangers, where a dictator shared an unfair amount of money—say, $1 out of $10—with a victim. The study participant would then have to decide how much of their own money, out of $5, they wanted to spend to redistribute funds from the dictator to the victim. Sure enough, the compassion meditation participants were more generous during the Redistribution Game. "They actually were kinder and they shared it more," Dr. Davidson told me about those who spent two weeks working out their compassion muscle.

The study also used functional magnetic resonance imaging (fMRI) to measure brain responses before and after the experiment. While in the scanner, participants viewed images that depicted human suffering, like a burn victim or crying child. Using their meditation practice, they generated feelings of compassion for these images. The researchers found that the participants who were the most altruistic after engaging in two weeks of daily compassion meditation exhibited the most changes in the brain during this part of the experiment. Specifically, they found that activity in the inferior parietal cortex, a region of the brain involved in empathy, had increased compared to before the experiment. Researchers also saw increased activity in the dorsolateral prefrontal cortex, an area of the brain associated with emotion regulation and positive emotions. The results of this study showed researchers that it could be possible to train the brain to be more compassionate.

There are two types of learning for the human brain. There is declarative learning, which is how the brain learns a new skill and acquires new information. We can learn about kindness, Dr. Davidson said, but that doesn't necessarily mean we will become more kind. Then there's procedural learning, which is the acquisition of habits and certain types of cognitive skills. Neuroscience shows that this kind of learning activates different brain circuits to experience real transformation in the brain. It can create habits.

"There are many possible applications of this type of training," Dr. Davidson explained in a media release about his research on compassion training.[7] "Compassion and kindness training in schools can help children learn to be attuned to their own emotions as well as those of others, which may decrease bullying. Compassion training also may benefit people who have social challenges such as social anxiety or antisocial behavior."

Three years later, in 2022, Dr. Fratantoni coauthored a study published in *Frontiers in Psychology* that scientifically demonstrated that helping others learn about kindness can build resilience in the brains of the givers.[8]

As many people experienced, the COVID-19 pandemic disrupted nearly all aspects of daily life. Parents of young children felt this very acutely as childcare centers and schools shut down. Despite this, parents still had to work from home with no paid sick leave or an equivalent to support them. Dr. Fratantoni and her colleagues recruited 38 parents of three- to five-year-old children to follow an online, self-paced kindness curriculum for four weeks. In the program, called *Kind Minds with Moozie,* a digital cow named Moozie shares creative exercises that parents can do with their kids to teach kindness. The exercises focus on four different themes of kindness: kindness to oneself, kindness to animals, kindness to others, and kindness to the earth. An example of a prompt is a parent asking the child, "How can we be kind to each other this morning? Let's ask Moozie!" And Moozie says, "Let's say good morning, family, and good morning, neighbors. Good morning, Moozie! Moo!"

To determine how kindness influenced the brain health of the parents and the kids, the researchers asked the parents to survey their own resilience and report on their children's empathy before and after the Moozie course. They found that the parents demonstrated more resilience and the preschoolers appeared to be more empathetic after the kindness training. "In times of stress, taking a moment to practice kindness for yourself and model it for your children can boost your own resilience and improve your child's prosocial behaviors," Dr. Fratantoni said, adding that people likely don't realize that kindness and altruism are must-haves for optimal brain health.[9] "It sounds like a nice [thing] to have, it's a fluffy thing, but I think the way we were wired and how we were created for connection, it's a have-to-have in order to really thrive," she said. One big takeaway from her study, Dr. Fratantoni said, is that practicing kindness for just a few minutes a day can help build resilience in the brain to better endure future stressors.

"No man is an island, / Entire of itself," poet John Donne famously wrote. "Every man is a piece of the continent, / A part of the main." Thinking about Dr. Davidson and his quest to make kindness and compassion as much of a habit as brushing your teeth, I wonder once again, is it possible to prioritize caring, kindness, and altruism in a culture built on scarcity? How can existing infrastructures change? Or does it really have to be this complicated, as some have previously said? To find answers, I hopped on a plane and traveled to another island.

. . .

For over a century, Western tourists have been drawn to the diverse archipelago islands situated a little over 2,000 miles from the state of California. This is in part thanks to how Hawaii has been marketed. While native Hawaiian culture is intricately connected to nature, storytelling, music, dance, and the native language, the state of Hawaii has often been simplified to attract tourists and packaged as a place for hedonistic fun in the sun, a destination full of the "aloha spirit."

In 2019, Hawaii saw more than 10 million tourists visit its shores, meaning that there were nearly 250,000 visitors in Hawaii on an average day.[10] Hawaii is home to a population of 1.42 million people.[11] I'll spare you the math, but this means that tourists outnumbered the local population almost every day of the year. It led to crowded and polluted beaches, bumper-to-bumper traffic on highways, and 90-minute wait times at restaurants. "I can't even take my kids to the beach on a weekend because it's so crazy," a local told the Associated Press as tourism peaked in the 2010s. There were hospitality worker shortages.[12] Native Hawaiians lost their homes to the construction of resorts and hospitality infrastructure. The uptick in tourism burdened locals. Something needed to change.

At the time, Chief Brand Officer Kalani Ka'anā'anā of the Hawai'i Tourism Authority (HTA), who has lived in Hawaii all his life and whose family spans five generations on the islands, was tasked with renewing the agency's strategic plan for tourism. Previously, the HTA's strategic plan was to increase tourism and tourist spending, and it worked. Each year between 2016 and 2019, visitor records topped each other. However, not everyone was happy with the growth, as Ka'anā'anā found out during his one-year "deep listening tour." For 365 days, Ka'anā'anā had honest conversations with Hawaiians across the islands through various town halls to understand what they wanted to change with regard to tourism in the future. "We heard time and time again, how do we continue to deepen the visitor industries' benefit to our community?" Ka'anā'anā told me in an interview. "How can we make them feel a part of something bigger than themselves, and to get them to give back and take a literal hand in making sure we can mālama Hawaii, or care for Hawaii?" People would say, we have all these people, all these hands. We have talent, experience, and tens of thousands of different skill sets visiting our islands every day. Hawaiians wanted to work with Ka'anā'anā to find a way for tourists to have a "less extractive" experience in Hawaii. "How do we create an infrastructure for them to be able to do so?" he said. Mālama, he explained, is a sister value to aloha. "It means 'to care for' in its most simple sense, but it's also a verb, it's something we do and that we actually take action on," he said.

Ka'anā'anā asked if he could play a little game with me. He was going to say the names of famous cities, and I had to share what came to mind. First he said the Vatican. Obviously, Catholicism and the Pope. Then he said Las Vegas. Casinos, glamorous evening shows, and the flashy lights of Las Vegas Boulevard. What happened when he said Hawaii? I envisioned surfers and a stunning beach.

"What we're trying to say is that the Hawaii that exists today and the Hawaii we envision in the future [are] very different," Kaʻanāʻanā said. Tourism, he said, is an important part of the human experience. Travel connects people with other cultures. It can bring our world closer together. "And we have a unique opportunity to share a little bit on our way of thinking about it," he said.

Following the input, the HTA created a plan to focus on regenerative tourism, a way to incentivize tourists to volunteer on their vacations. For example, hotels could offer a free night's stay or a discount if tourists signed up for a volunteer event with a local organization that needed extra hands. The idea, Kaʻanāʻanā said, is that residents live to be stewards of their land and community. Could tourists adopt a similar mindset on their vacations, and perhaps, even take that mindset home with them? In January 2020, the plan was approved, but the rollout faced a major barrier: COVID-19. In February 2020, the world shut down and travel to Hawaii came to a halt. Kaʻanāʻanā and his colleagues went back to the drawing board to figure out if this idea of regenerative tourism could be sustained when Hawaii reopened. Today, the kinks are still being worked out. As Kaʻanāʻanā said, Hawaii is playing the long game with this. "Regenerative tourism isn't just a new buzzword," he emphasized. It's not sustainable tourism. It's not responsible for tourism. "It's more about how do we disrupt the systems that created the present conditions that we have, and really create a new paradigm for what travel and hospitality could be?" he explained.

I met Kaʻanāʻanā nearly two weeks after the island of Maui experienced the deadliest U.S. wildfire in more than a century.

Nearly a week and a half before our meeting, a fire in Lahaina, formerly the capital of the Hawaiian Kingdom, burned more than 2,700 structures. At the time of our interview, hundreds of people were still missing as the death toll kept rising. What exactly sparked the wildfire remained unknown at the time of our interview. But experts were already speculating that a mix of Maui's dry weather, low humidity, and strong winds from Hurricane Dora 700 miles to the south likely contributed to the destruction. In other words, climate change had likely contributed to the severity of the fire. Images and stories showed people desperately jumping into the ocean to save themselves and their loved ones as flames burned down their homes. At the same time, stories surfaced about tourists carrying on with their vacations as people were forced to reckon with the ashes of their lives. True to the nature of a wildfire disaster, stories also began to surface about how the community was banding together to make "Maui strong."

As I left my interview with Kaʻanāʻanā, we talked about how volunteerism is a way to build resilience in the wake of a crisis. Cultural and political differences are cast aside when a disaster like a wildfire in Maui strikes, he said. But practicing mālama in noncrisis times has contributed to the state's resilience as well. Indeed, Hawaii has the second-lowest rate of gun deaths in all of America. The state is less politically divisive than other parts of the United States, despite the constant reminders of colonialism, war, and tourism on various islands.

As we walked together, he brought me over to an exhibit about the healing stones of Kapaemahu.[13] According to ancient Hawaiian oral history, around the year 1400 four healers left their home in Tahiti and sailed to Hawaii. When they arrived, native Hawaiians excitedly greeted them. The healers gave much to Hawaii, as they could heal by touch and diagnose illness, and displayed great spiritual power. Healing, to Hawaiians, is not just about treating a symptom. It's about looking at the full picture of a human being, of a system. The impressive powers of the healers were in part due to their being māhū, meaning they had both male and female spirit in their hearts and minds. Upon the healers' departure from Hawaii, they instructed people to fetch four gigantic stones to be placed as permanent reminders of their healing capabilities and all the good they had done. After a month-long ceremony, they transferred their powers to the stones. After using half-male, half-female idols, representing the feminine and masculine in everyone regardless of biological sex, to facilitate the transfer, the four healers disappeared.

Anywhere else, Kaʻanāʻanā said, this exhibit would be "political" or "controversial." But not today, not in Hawaii.

. . .

On December 7, 1941, just before 8 a.m. local time, the Imperial Japanese Navy Air Service launched a surprise military strike of torpedo planes, level bombers, and dive-bombers over the U.S. naval base at Pearl Harbor.[14] In two waves of attacks, four ships sank, killing 2,403 people and wounding another 1,178. To this day, Pearl Harbor remains the deadliest event ever recorded in Hawaiian history. But the losses of that day extended beyond Hawaii and the United States. First, the United States declared war on Japan. Then, Germany and Italy, who allied with Japan, declared war on the United States. America officially entered World War II.

Today, the USS *Arizona*, the ship where half of those who died at Pearl Harbor were located, lies below an American flag flying above the water. In the same water, nearly parallel to the sunken *Arizona*, is another naval ship that wasn't there on that fateful day but is responsible for ending the horrific war that followed: the USS *Missouri*. Nearly five years after the Pearl Harbor attack, General Douglas MacArthur stood on the teak decks of that naval ship and called for justice, tolerance, and rebuilding. World War II finally came to an end thanks to the events that happened on that ship. When the USS *Missouri* was taken out of service and needed a permanent place to dock, the state of Hawaii said that since they had the beginning of the war, commemorated by the USS *Arizona* Memorial, why not park the USS *Missouri* next to it to showcase the bookend of the war.[15]

Curious to learn more about how Hawaii was trying to shift tourism from aloha to mālama, I drove to the Pearl Harbor entrance and took the shuttle to Ford Island to meet a man named Keven Williamson. Standing tall in a green polo shirt and a beige sun hat, Williamson had a personal history with Pearl Harbor. In the 1970s, he served on a frigate ship there. A winding career led him to where he is today: the USS *Missouri*'s volunteer director for the past 15 years.

From the start of our meeting, Williamson is eager to show me the ship and his crew, who are cleaning its living quarters, where every inch of space is optimized for utility. Three bunk beds are stacked on top of each other from floor to ceiling. In addition to being a destination for tourists and history buffs, the decommissioned ship is in constant need of volunteers to keep it operating properly. "You have to remember, this ship is like a small town," he said. "Whatever you have in a small town, you have here." Except this small town no longer has a full staff. When it was in commission, there were 2,500 sailors maintaining the ship. Today, there are fewer than 25 maintenance staff. For this reason, it's now part of Hawaii's mālama regenerative "voluntourism" program. And under Williamson's command, there's no shortage of work to be done.

When tourists volunteer, Williamson said he usually has them simultaneously clean and enjoy the tour route. With visitors from all over, the need to clean is constant. At the same time, he doesn't want to compromise on fun and learning. He tells volunteers to take the tour route for free, but simply wipe clean where they think people will place their hands. While they're doing it, Williamson said, they're looking around and learning about the ship. "It's excellent for us," Williamson said. "And they are so excited about it."

The impact has been instantly noticeable to Williamson and the ship's crew. When COVID-19 hit, they lost a lot of regular volunteers. But the mālama program has helped recruit and supply more volunteers as tourism has bounced back. It's a win-win situation, he said, adding that people leave feeling better than when they arrived. Williamson said he suspects it's a more impactful experience for people compared to when they just visit the ship and take the tour. In part, it's because they have a sense of giving back and preserving history. "Not only do people who come get something out of it, but we get something," he said. "You are helping us do our jobs to preserve history and I think that's very important to people."

It's not just tourists, but companies, too. For example, he said that workers from General Electric, who helped contribute to the original construction of the ship, still come and volunteer on the ship. They also get a lot of Japanese tourists, students, and people who previously served in the Japanese Navy. Afterward, they feel so proud of their work, Williamson said.

As I walked around the main deck after speaking with Williamson, I imagined what life must have been like when the ship was in its heyday—buzzing, a little cramped, and perhaps even a little scary with so much ammunition and firearms on board (eerily reminding me of the Halifax disaster). Not exactly my cup of tea, but I did feel a slight pang of envy for the camaraderie they must have felt on board. To know that there was always someone to talk to, and to always be working toward a common goal. To always have an opportunity to be of service and be connected to something bigger than themselves.

As the wind nearly blew my hat off once again, my attention was drawn to a photo of a kamikaze who crashed on the starboard side of the main deck of the ship. The story goes that a kamikaze aircraft crashed on the USS *Missouri* when it was anchored northeast of Okinawa. With part of the plane's wreckage falling into the sea, the pilot's dead body fell onto the main deck. Per the orders of Captain William M. Callaghan, a chaplain on board conducted a military funeral for him the next day, when they prepared a Japanese flag. The story is seen as an example of the importance of compassion even in the midst of war. As I thought a little deeper about its symbolism, I couldn't help but optimistically think that it provided a little resilience to the crew that day, especially to anyone who felt internally conflicted about the war. In the middle of so much death and bloodshed, perhaps they felt slightly less depressed and daunted by catching a little glimpse of the humanity that remained.

There are many paradoxes and nuances regarding the USS *Missouri*. Not only was it a purveyor of peace, but it was also responsible for death. Following the World War II surrender, the USS *Missouri* went on to participate in the Korean War, carrying out naval raids on North Korea. During the Gulf War, the USS *Missouri* fired a Tomahawk missile at Baghdad. Yet people from all around the world come to volunteer and preserve it. For these reasons, and more, I asked Williamson what the USS *Missouri* symbolizes today. He responded, "Strength." But to me, it's not a stereotypical masculine strength he was talking about, despite the fact that the ship's hull is protected by massively thick steel armor. Instead, it's the strength that comes from working together, despite one's differences. Today that takes place in the form of volunteering. In the navy, he said, a commonly used phrase was "the captain is only as good as the worst sailor on the ship." Perhaps this is a lesson that can be applied to how everyone lives their everyday lives. On board a ship, and in life, we're better when we help each other be the best we can be. "This is what it's all about," he said. "It's about working together."

7 *The Mālama Mindset*

To my right, the turquoise blue ocean sparkled. To my left, green mountains and their jagged edges stood tall. As my car twisted and turned between the two, I slowly veered into the tropical rainforest of Waimea Valley, a popular destination for tourists to hike and find solace in nature.

Before it became known as a popular tourist destination, the valley held historical significance for native Hawaiians, as it's home to many ancient structures. But like much of Hawaii, Waimea Valley has been deeply affected by capitalism and colonialism.[1] A brief overview goes like this: prior to 1700, a Hawaiian chief ruled Waimea Valley, which was a stretch of land extending from the mountains to the sea. Alone, it had enough resources for people to survive and live off the land. After 1700, a famous priest named Kaʻōpulupulu, who is believed to have predicted the arrival of foreigners in Hawaii, lived in the valley. Once Kamehameha the Great conquered Oahu in 1795, he gave Waimea Valley to a trusted advisor named Hewahewa Nui. During this time, foreign influences were changing Hawaii, and its traditional system of laws (kapu) began to fall. In 1837, Nui gave the valley to his granddaughter Paʻalua. Sadly, after the passing of the Mahele Land Redistribution

Act in 1848, she had to give up half her ownership of Waimea Valley. Then, in 1884, high debt forced her to lease what land she had left. In the early 1900s, Waimea Valley ranchers used the valley, and the U.S. military occupied it. All of this is to say that its history has greatly affected Waimea Valley's topography and landscape.

But I don't know to what extent this land has been impacted when I drive in. Instead, I'm more focused on finding much-needed relief from the blazing sun. Initially, I'm struck by how quiet and serene the valley feels compared to the rest of the busy North Shore. How lush and green the plants are around me. A cool ocean breeze harmonizes with bird chirping and I can feel my nervous system slowing down after being activated by my being a few minutes late. Michael Herrera, the volunteer coordinator of the nonprofit Hiʻipaka LLC, which manages conservation efforts in Waimea Valley, ushers me through the greenhouse to where a group of eight more volunteers are waiting for me. He asks if there's anything I need to do before we hike 45 minutes up to the conservation site where they've been planting native koa trees for the past few years.

The hike begins with a chant dedicated to the Hawaiian god Lono, who symbolizes agriculture, fertility of the land, and rain—poignant in the wake of the Maui wildfires and another, less heard of, wildfire that occurred on Oahu the day before. Together we clap, call, and receive. One of our guides notes that the chant is customary before every hike. In ancient Hawaii, he says, people utilized chants to let their presence be known to the valley. It was a greeting of sorts. A simple act of kindness. I thought back to how, in Paris, a Frenchman stereotypically complained to me about how Americans are so rude. He said I should always say "bonjour" before asking for something. It's important to greet and recognize the presence of another person, instead of just approaching them with a request.

I don't know anyone in the volunteer group personally, but it's not hard to feel like I'm already part of *something*. I feel a sense of purpose, like I'm on an important mission. McKenzie Latimer, a conservation specialist at Waimea Valley, says that everything we will do today will not only help the koa trees, but the entire valley. "It affects everything that goes up to the stream," she said. "It's all a cycle." It's just like how the early Hawaiians used the wood of the koa trees to make canoes, spears, and paddles, she says.

In Hawaiian, *koa* means "brave, bold, and fearless."[2] When given a chance, a koa tree can grow five feet per year for the first five years. At its peak, it can reach nearly 115 feet at high elevations. However, in a forest like Waimea Valley, which is now full of invasive species, thanks in part to its fraught history, their chances of reaching their full potential are low. This is precisely why we are here today: to spend the day pulling, sawing, and snipping invasive plants, to give the fresh, vulnerable koas a fighting chance. This program is another part of the Mālama Hawai'i initiative to get tourists to give back during their Hawaiian vacations.

The trail to our destination starts on a fairly flat surface. As an experienced-ish hiker, I'm not so much intimidated by the hike as I am by the heat. It's probably almost 90 degrees Fahrenheit out, and I'm sweating a lot. I try to stay present and observe the foliage around me. I admire its greenery and think about how Hawaii has so many beautiful plants and trees—until Latimer points out that, 15 minutes into our hike, we have yet to see a native Hawaiian plant. All the ones I've been admiring are invasive.

Latimer says the strawberry guava plant, in particular, thrives in Waimea Valley. In fact, it poses a massive threat to the future of the koas. All across Hawaii, researchers have stated that the greatest threat to native Hawaiian forests aren't wildfires, volcano eruptions, or even feral cats, but these medium-size strawberry guava trees. Native to Brazil, strawberry guavas look like engorged cherry bulbs emerging from dark green leaves. The contrast in their colors reveals a bit about both their sweetness and their menace. Once Latimer points out the strawberry guava, I realize the extent to which they've successfully taken over. They're everywhere. She stops in front of one, pulls a bulb, and asks if I want to take a bite. It feels a bit like a test. I'm getting Adam and Eve vibes. What if I enjoy this forbidden fruit? I hesitate, but agree. And it turns out I do like it. One bite in and I'm hooked, tempted to google if I can find these in Whole Foods on the mainland. It's the perfect combination of sweet and tart, leaving me nostalgic for the strawberry Jolly Ranchers I indulged in as a kid. While I feel slightly guilty that our assignment is to commit herbicide on these tasty plants, I'm not alone. Latimer admits she likes them, too.

Certainly their taste is part of the appeal. Historians say the first strawberry guava plant arrived in Hawaii in 1825.[3] While the intent behind bringing them over was likely harmless, their presence has contributed to Hawaii becoming the endangered species capital of the world.[4] This means that on the Hawaiian islands, there are more endangered species per square mile than in any other place on the planet. In the U.S., Hawaii is home to 44 percent of the nation's endangered and threatened plant species. With its seeds passing through the digestive tracts of birds, livestock, and wild pigs, the strawberry guava plant has been propagated around the islands by wildlife, which have unknowingly collaborated in the plant's ruthless takeover. The fact that Hawaii bolsters an all-year growing season doesn't help the situation either, as it expedites the growth of these plants. Sometimes referred to as the "sweet invader," strawberry guava plants are quick to shade out native plants that are part of complex native ecosystems. It's estimated that in some regions of Hawaii, strawberry guava plants have reduced groundwater recharge by 85 million gallons a day as a result of this action. Since native Hawaiian plants have evolved to have very few defense mechanisms, they need conservationists and volunteers like us to get them out of this mess.

Finally, about halfway through the hike, we walk past a native Hawaiian plant. Latimer stops us to point it out and put into perspective how few of them there have been. Ironically, it's next to coralberry, an Australian Christmas holly that clearly has no business being in Hawaii. When I ask Latimer what is being done island-wide to battle invasive plants and protect endangered plants and animals, she says there is no all-hands-on-deck statewide plan. Instead, it's more like a bunch of different conservation groups doing what they can. It's almost as if it's become too big a problem for the state to solve.

Reflecting on an issue that the entire country is grappling with, Michael Herrera, the volunteer coordinator, also points out that conservation groups who are trying to save native Hawaiian species are underfunded, which means everyone is doing what they can with strained resources. As we transition from flat ground to switchbacks, I ask what's at stake if this problem isn't solved. What will happen to Hawaii, the endangered species capital of the world, if the invasive species win?

"I think the biggest threat right now is the loss of entire species," Latimer said. Losing native plant and animal species could lead to immeasurable amounts of ecological damage and disrupt an ecosystem that's struggling to function as it's meant to in order to thrive, she added. There's also an emotional connection between Hawaiians and many of the plants that are on the verge of extinction. For many, losing a plant species is like losing a family member.

After trekking up an exhausting incline, through a noticeable drier forest populated by pine trees, we finally arrive at the site where hundreds of acres are home to hundreds of koa trees.

The ocean plays peekaboo with our new location at the top of the ridge, depending on where we are standing on the site. We are given a moment to catch our breath before we get to work. As Latimer said, the mission of the day is to clear space around the baby koa trees and save them from the strawberry guava plants that are plotting to suffocate them. And now I can see why. The baby koas look so young and vulnerable. With the trees protected by fencing to keep away the invasive pigs roaming the forest, it appears that the organization's hard work is paying off. Latimer points to the baby koa trees and says this is what a native Hawaiian forest is supposed to look like. It's less busy, and more calm. Less wet, more dry. We are told to work in pairs, using either a saw or heavy-duty garden scissors to cut away the invasive plants around the baby koa trees. Once they're snipped down to their roots, we have to dribble a little herbicide on them, or else we risk the chance of them growing back. They're that tenacious.

Latimer says native Hawaiian plants like koa trees are vulnerable because they didn't evolve to have strong defense mechanisms, like thorns or a yucky taste. It's in their nature to grow at a slower pace than mainland plants. Despite mandated inspections at the Agricultural Inspection Counter when agricultural products arrive in Hawaii, there are lots of insidious ways for invasive plants to find their way into the state and even up to this sacred site. For instance, a seed can get on someone's shoe during a hike on the mainland and end up right where I'm standing. Restoration of the koa trees is good for the entire ecosystem, and the effects are noticeably visible. Since volunteers have been removing the invasive plants and replanting koa trees, the soil is more fertile and full of iron where their efforts have been successful. Latimer points out how the soil is visibly more red at the conservation site.

I'm lucky enough to partner with Latimer, who identifies the first koa we are assigned to save. It's about four feet tall and has a trunk that I can wrap my thumb and index finger around. The color of its wood is slightly darker than sand, but a bit lighter than an almond. However, at its roots, baby strawberry guava plants are clawing their way around the baby koa to the point of near strangulation. Latimer lifts the plant's slinky, snakelike branches and takes the first snip. I feel a small sense of relief as a branch falls to the ground. She hands me the scissors and I snip a few more until all that's left is a constellation of tiny trunks. Latimer proceeds to pour the green herbicide over the decapitated bouquet, and our job is done. On to the next koa.

Latimer holds a handful of strawberry leaf branches and I snip them down to their roots. Herbicide is committed once again. When I ask if it's hard to train new rounds of people, sometimes tourists, multiple times a week, she says yes, but it's still worth it. Even if it's not the most efficient strategy, Latimer says educating tourists about the issue of invasive plants on Hawaiian islands is just as important as, if not more important than, eradicating them. The goal is that by educating people, it can also inspire them to change their habits at home. "Just simple things like planting native plants, it can have a bigger impact than anything we can do alone," she said. "It's a trickle effect." But having more people helps. "It's a dual effect as well," she said, noting that not many of them work at the nonprofit full-time.

Latimer didn't start her career with a goal of being a conservationist and spending her days killing strawberry guava plants. She actually worked in the mining industry in Canada, but came to the realization that she wanted to help the environment. Hailing from generations of loggers and foresters, she sees herself keeping up the legacy of her family's work, but in a different way—one that preserves forests rather than cuts them down.

After we saved our third koa tree, it was time for lunch. All eight of us gathered under a nonnative tree and freely talked about Hawaii and climate change. One volunteer said she was worried about flooding in the future, in addition to wildfires. Another claimed that Oahu has one of the largest landslides in the solar system. I later confirmed that the Nuʻuanu slide lies off the northeast coast and is indeed one of the largest landslides on earth. Another volunteer pointed out a bird that was flying by, and someone chimed in and said that there is a spiritual significance to them. I think about everything I'm learning and how it wouldn't be possible if I wasn't with this unique group of people at this very moment in time. Every person there has something to offer. There is something very human about engaging in a casual conversation with strangers who quickly become friends. There is also something different about this experience from talking to strangers on the street or in a restaurant, which requires another level of vulnerability. By volunteering in a group, together, we are already bound by the task at hand: saving baby koa trees. We don't have to get to know each other or feel as if we are interrupting someone, because we came together to accomplish something with a sense of purpose. Initially, it's this common purpose that makes us more willing to be open and share with each other. Some people even offer to share their snacks with the group. It also occurs to me that I'm conversing with people who are different from me in age and background. Sure, anyone here could have decided to hike Waimea Valley on their own with their friends and family, but there is a unique dynamic to sharing this experience with people whom I otherwise would have probably just waved to on the trail and passed by.

After lunch, I stayed behind for a few minutes to chat with Herrera. I asked him about the impact of what we were doing that day. Is it helpful to have tourists volunteer? In a situation where there is very little training, and you're dealing with people who aren't exactly in work mode? As a volunteer coordinator, he said volunteering in a situation like today's is how strong communities are built. It's how the people of the world can strengthen connections between each other. "The more you meet people, the more you realize everybody is the same," Herrera said. "As people learn more about other people and share time with other people from different places, it builds our acceptance of diversity." It also builds relationships among people who might not otherwise cross paths, he added. "It's a giant root system that's growing and creating something that more can come from," he said, excusing the cheesiness of the tree metaphor.

On a more quantitative level, volunteers can make a big impact on Waimea Valley conservation efforts, he said. There are 1,800 acres and over 50 gardens that need to be cared for, which means they need all the help they can get when they're up against the ruthless strawberry guava. The nonprofit has a small staff. Then there's the crisis of climate change: as temperatures increase in Hawaii and rainfall becomes less frequent, many native plants are not well positioned to adapt to the changing ecosystem, especially when competing against invasive species that are more adaptable. Already, increasing temperatures and more frequent droughts are causing dramatic declines among native plant species, like the Haleakalā silversword. However, by volunteering in Waimea Valley, we are helping the native koa trees build resilience against these ongoing threats.

People often think climate change is too big, too abstract, that it's a future crisis that an individual can't help address with their individual efforts. "It seems daunting, and it is, but at the same time it's only through individual actions that we got here," Herrera said. "So why can't we reverse, repair, and mitigate the effects of climate change as individuals?" Individual action can make a difference, and at the same time, individuals working together can make even more of a difference. Collective action, he said, is incredibly powerful. "We have a small group of people here, yet for every circle we make around a koa tree, we are giving that tree a higher likelihood of surviving," he said. "The more individuals that come out to do that, the more trees are going to survive."

As a volunteer coordinator, he often hears how people are more willing to donate their money than time. Free time in our society is limited and rare. Yet, he sees again and again that volunteers get a lot out of giving their time, more so than by donating money. The reward is often greater than that provided by participating in a favorite recreational activity.

Just as I'm about to ask another question, Latimer comes over to check on us. "I wanted to make sure you're okay," she says, admitting she's always in "safety mode." Certainly, it's not too hard to get lost up here. Considering all the edges and cliffs, one could easily slide down a hill and get hurt. It occurs to me that this is another benefit of group volunteering. There are plenty of horror stories of hikers getting lost while on vacation. But volunteering to save koa trees in Waimea Valley strikes me as being a win-win scenario. I'm helping preserve the forest and meeting new people, and I have a built-in community to look after me.

I return to the conservation site with Latimer to save a few more koa trees. Snip, saw, and pour the herbicide. During the second half of the day, Latimer and I are in flow and chatting some more like we're old friends. While we may never meet again, I will remember so much that I learned from our time together, like the fact that she shares a birthday with my daughter. At the end of the day, another conservationist, whom I didn't get the pleasure of knowing, thanks all the volunteers for their time. He emphasizes that we really helped preserve the land today. We saved about a dozen koa trees. We did our part in Mālama Hawai'i. In reflecting on how I'm feeling, I'm aware that I'm experiencing what is often referred to as the *helper's high*, a term coined by Allan Luks in the late 1980s that describes the powerful and positive feelings a person experiences after helping others.[5] Luks's own research has found that people who are "regular helpers" are 10 times more likely to be in good health compared to people who don't volunteer. His research found it's because volunteering reduces the body's stress through the release of endorphins, just like when a person does exercise.[6]

Luks isn't the only researcher who has found that people who volunteer regularly are in better health than those who don't. Not only do they have better physical and mental health, but more science is emerging to suggest that volunteering in groups, specifically, can help people build more resilience to illness and disease later in life. For example, some scientists think volunteering could serve as an intervention for all older adults against Alzheimer's disease.

In fact, in research presented at the Alzheimer's Association International Conference in 2023, epidemiology doctoral student

Yi Lor, of the University of California, Davis, looked at the volunteering habits of 2,476 older adults.[7] Forty-three percent of the participants reported volunteering in the past year. He said he found that those who volunteered scored better on tests of executive function and verbal episodic memory. When the researcher adjusted for age, sex, education, income, practice effects, and interview mode, they still found this to be true. When looking at frequency of volunteering, the researchers said that those who volunteered several times per week had the highest levels of executive function. Volunteering, Lor found, keeps a person's brain engaged. It also helps people socialize and lowers stress.

In a separate study, Dr. Eric Kim, an assistant professor of psychology at the University of British Columbia, looked at a sample of more than 7,000 adults over the age of 50.[8] He found that those who volunteered had 38 percent fewer hospital visits than those who didn't. They were also more likely to engage in preventive health care measures than nonvolunteers.

When I first interviewed Dr. Kim about this study in 2017, he revealed he was a bit cautious to say volunteering was the key to living a long and healthy life. He worried it was a bit of a chicken-and-egg kind of situation. It's hard to know if healthier people are attracted to volunteering, or if volunteering makes someone healthier. Five years later, I asked him if he still felt cautious. He didn't. "Since then, there's been more research that we've done and others have done as well, and it does appear like volunteering, even in randomized control studies, has great health behavior and health outcomes," he told me, adding that opportunities to volunteer need to be prioritized by our society.

In our world, we've all met older adults who appear to be younger. Maybe they physically look younger, or maybe they are more active than other people their age. When it comes to aging, humans usually define it by their chronological age, as in years lived since their dates of birth. But there's also a person's epigenetic clock, which is a person's age based on specific biomarkers determined by the biochemistry of cells, tissues, and organ systems. This is all being studied in an emerging field of science called epigenetics, where some researchers are trying to figure out how to reverse aging. Dr. Kim, and other scientists, believe that volunteering in old age could be one way to do that. In fact, in one of Dr. Kim's studies, he found that volunteering was associated with reduced epigenetic age acceleration in 6 of 13 epigenetic clocks.

"One of the findings that many labs keep finding over and over again is that more volunteering is associated with reduced risk of mortality," he told me. When I asked if he thought volunteering should be used as a health intervention to build resilience on a personal health level and societal level, he said absolutely. He added that the government should create more government-sponsored programs to provide funds for volunteering. It could be as simple as helping with transportation to get to the volunteering opportunity, to incentivize and lower barriers for more people to volunteer.

Back in California, I thought more about how in the face of an activity that is extremely self-centered, like a vacation, a volunteer activity can be the most memorable part of it. Science says a day hiking in nature would have certainly been good for my health. Like I said, I've done many hikes in my life, but not like the one in Waimea Valley. I can tell that the benefits I feel have been amplified and will be long-lasting, that this will be an experience I'll remember for years to come. Acts of kindness are good for our well-being and health. They have the potential to build resilience in our society and act as bridges from a state of loneliness to feeling connected. In the wake of a crisis, an act of kindness has the power to regulate our nervous systems and make us feel safe. But maybe the key to building true resilience on both an individual health level and societal health level is to not only create systemic caring and make a habit out of acts of kindness, but also to create more accessible opportunities for people to routinely volunteer in groups together.

I find that the lesson to be learned from the Mālama Hawai‘i program is that there is value in having a mālama mindset—in other words, for people to take on a cultural change. And making a cultural habit out of group volunteering is part of that. But how can we make it so volunteering isn't something that's done only once or twice a year? And what can the government do to promote, support, and sustain it?

8   *Volunteering for All*

Research shows that 90 percent of Americans want to volunteer their time, meaning they want to spend their time helping improve their community without being paid, but only one in four actually does.[1] When asked why they don't, people often say they don't have enough time. Volunteer schedules are inflexible. People don't have enough information, or they were never asked to volunteer. Today, volunteering is very much an activity that is accessible only to people with resources and privilege. Oftentimes, for people who have the resources to give away their time to help others without being paid, it's something that is used to improve a résumé.

The very concept of volunteering was created through a capitalist lens. To volunteer implies that people are donating their time to help others and will receive no monetary reward. There are potential health benefits, but since the benefits aren't making a profit, volunteering appears to be less accessible in our modern society. To volunteer is a mutual act of giving and receiving, and to position it as an activity that could easily be replaced by a more "productive" one undermines its impact. But historically, it's volunteers who have led systemic change. The word *volunteer* itself is derived from the French word *volontaire*, which was used in the

1600s to describe someone who gave themselves up for military service.[2] During the Revolutionary War, citizens volunteered to boycott British imports. There were also "minutemen" who made up a volunteer militia. Many established organizations started as volunteer-based movements, like the YMCA, the American Red Cross, and the United Way. In 1736, Benjamin Franklin developed the first volunteer firehouse, an idea that lives on today.

But more scientists and public health experts are making the case that volunteering should, and could, be a health intervention. As Robert Putnam, author of *Bowling Alone*, speculated, some of the same reasons that have led to the decline of civic engagement are likely contributing to the barriers people face when trying to access volunteering. Is there a newer, more accessible way for the government to promote volunteering? As we saw with the social prescribing movement, it's just in the beginning stages of taking off in the United States. Some doctors are open to prescribing volunteering. But what can the government do to help, aside from the obvious—like promoting paid time off, incentivizing employers to allow their employees to take time off to volunteer, and providing more livable wages and social support to people?

In 2008, Josh Fryday joined the U.S. Navy. As an American citizen watching the horrors of the War on Terrorism go on for so long, he had a gnawing sense that many Americans didn't have enough invested in it.

It was his experience in the military that led him to a realization that would change the course of his career: that having service experiences, where a person is forced to work with people from different backgrounds and with different perspectives, all while having a sense of purpose upon admission to the military, was really valuable. In fact, he thought it could be the antidote to the political polarization that was increasingly rising. After his tenure in the navy, he went on to start a national organization to fight climate change. Then, he became the mayor of Novato, a town in the Bay Area, where he started a program called Reimagining Citizenship with Dominican University. In this program, undergraduate college students could serve the city of Novato while earning a degree from Dominican University.[3] The city of Novato provided students with a $10,000 stipend for work to be completed over two consecutive summers, and Dominican University provided up to $100,000 in scholarship funding.

"It was a first attempt of how we as a city, as a community, value service and then send a message to our young people: if you're willing to serve, we're going to help you with your career," Fryday told me in an interview.

The program was a lot like the GI Bill, he elaborated, but one that had a modern spin on it and focused on community service, which varied from helping elementary school kids with their homework, to helping adults acquire some sort of professional license, to volunteering in the parks and recreation department.

When California governor Gavin Newsom took office in 2019, he appointed Fryday as the state's chief service officer. Since then, Fryday has been tasked with the goal of making volunteering more accessible to Californians. "How do we create a culture where people are expected to serve, help each other, and care about each other, but then [are] also given the opportunity?" he said, noting that our current culture doesn't make volunteering that easy. "We've created a culture where the values that we place as a society and government are on making money and taking care of yourself."

How do we create such a culture? Since being appointed, Fryday has invested in what he calls "building the civic infrastructure" by offering a handful of programs as part of the California Service Corps. The College Corps, for example, is loosely modeled on Reimagining Citizenship. In this program, each year nearly 10,000 college students across the state are given the opportunity to volunteer with community-based organizations working in three priority areas: K-12 education, climate action, and food insecurity. Through this volunteer work, each fellow earns $10,000. The idea is instead of working a job that the student doesn't feel has purpose or meaning, they can volunteer in their community and receive a stipend that will go toward their education. Fryday and his team also hope this will tackle another crisis: student loan debt. Another program is the Youth Jobs Corps, which prioritizes giving jobs to young people between the ages of 16 and 30 who don't have access to traditional career-building resources. This could mean someone who is low income, is transitioning in and out of foster care, or simply has difficulty finding employment. Again, in this program fellows work with community-based organizations that are already making an impact at a local level, while making at least minimum wage provided by the state. And then there's the California Climate Action Corps, which is the first state-level climate service corps program in the United States. In this program, fellows receive a $23,294 living allowance (before taxes) spread evenly over 7.5 months, which is the time they serve. As fellows, they clock in over 1,200 hours, with a focus on engaging their communities in climate action. They also receive professional development opportunities, such as the chance to receive climate-related certifications. Supplemental Nutrition Assistance Program (SNAP) benefits are given to those who are eligible, in addition to health insurance, forbearance on existing qualifying student loans, and childcare assistance for those who qualify. The goal of the Climate Action Corps is to support people mobilizing climate action in their communities. California is one of the states where climate change is most tangible, from snow melting earlier in the spring, to months-long droughts, to the increase in frequency and severity of wildfires. Fryday says Californians want to be part of the movement to build resilience in the face of these changes. Now he's creating the infrastructure to do just that.

Fryday and I met in Vallejo, California, on a warm spring day. He was in town from Sacramento to meet President Joe Biden at an event highlighting the state's Climate Action Corps, which Biden would later announce he was recreating at a national level.

This desire to take action against climate change can be observed in the California programs' popularity. In 2022, the Climate Action Corps received five times as many applications as there were slots available. The College Corps had three times the number of applicants as slots. Californians, Fryday emphasized, are "desperate" to connect with their communities. They want to take action and be part of change. "The challenge for us is how do we create enough opportunities for them," he said.

When asked what barriers Californians face to volunteering, he responded that he doesn't think the barriers are unique to California. First, he said there is a lack of infrastructure to plug into easily, which is what the California Service Corps is working on addressing. Another barrier is the culture. Fryday says it's clear there is loneliness in California, like in the rest of the country, because our culture pushes us to disconnect from each other rather than connect.

Another program that Fryday and his team are working on is the "neighbor to neighbor" program, which he hopes will encourage people to connect with their neighborhoods. The idea is that shared experiences, where people connect by being of service to their community, will strengthen social ties and help communities be better prepared for future disasters. "The reality is, and I think this is what we learned in COVID-19, is that in a disaster, the government can't be everywhere," he said. Then, in the event of a future disaster, like another pandemic or wildfire, communities will already have the infrastructure to band together and help each other. It could be as simple as through an email list. It could be funding a community garden or holding meet-your-neighbor events. Another benefit could be an improvement in people's health, which he said is deteriorating in part due to the loneliness epidemic. "What we're trying to do is create meaningful opportunities, authentic opportunities for people to connect where it's not just over a screen," he said.

As I'm talking to Fryday, I think back to Dr. Eric Kim's research on how volunteers have longer and healthier lives. When I asked Dr. Kim what it is about volunteering that contributes to longevity, he said that through hundreds of interviews he's conducted with retirees, he found it's because volunteering gives people a sense of purpose and meaning. A feeling of giving back to society is critical to aging in good health. But why wait until retirement to volunteer? For people who are still working, Fryday said, it's critical to shift the tone of and motivation for volunteering. Housing is a fundamental right. Food is a fundamental right. Community should be one, too, he said. "It's fundamental to living a meaningful life," he said. "It's fundamental to happiness."

Humans evolved to live in a community and collaborate with each other, he said. It's how we've survived, and we need it to survive as a species. "If you're on an iPhone all day long, or you don't know any of your neighbors' names, you're not collaborating with anybody," he said. "It's antihuman."

The work Fryday has done has made an impact on California, too. I later asked Fryday's office for data on each program. Thanks to the California Climate Action Corps, 11,068 trees have been planted, 191,262 trees have been maintained, and 20,946 trees have been donated. His office estimates that, in total, 28,884 volunteers engaged for 103,627 hours of climate action in 2023, and 5.2 million pounds of recovered food was distributed to people in need.

In a way, a program like this is very Roosevelt-esque and similar to the Civilian Conservation Corps (CCC), which was created by President Franklin D. Roosevelt in the 1930s. During that era, millions of young men served terms of 6 to 18 months in Roosevelt's so-called "Tree Army." Together, they helped restore national parks and the economy, all while supporting themselves and their families. While this New Deal program is often idealized, there were major flaws—like segregation, and the fact that only men were allowed to join.[4] Perhaps California really can reimagine a way to replicate the CCC today, but in a way that is accessible to everyone.

•

On a Tuesday afternoon, the next generation of California Service Corps members gathered at Fresno's City Hall.[5] The event was one of many taking place across California for its first statewide pledge ceremony for its newest members. Many people in attendance, and those participating via livestream across the state, were about to take a pledge to become fellows in one of the California Service Corps programs. But before they did, we were lucky enough to hear the stories of previous fellows, like Crystal Navarro, an AmeriCorps California member. Standing in front of the crowd, holding a microphone, Navarro introduced herself. She grew up in Fresno and was a graduate student at Fresno State University, where she was studying social work. As a student who had personally experienced incarceration, she got involved with the California Service Corps as a California justice leader in the Youth Jobs Corps. "The program helps individuals like myself complete our academic goals by providing mentoring, tutoring, cost of books, and they also give us basic needs grants," Navarro said. "More importantly, they treat their students like family." Thanks to the program, Navarro said she found herself surrounded by great role models and a strong support system. She grew in ways she could have never imagined. "For the first time in a really long time I felt included, accepted, and I had a strong sense of belonging," she said, as she held back tears. As a California justice leader, Navarro also found that her story wasn't unique and that there were others with similar stories.

"They founded this program on the belief that people can change," Navarro said. "I am proud to say that I have supported numerous young people with college matriculation and I have supported them at navigating systems and removing barriers that they once faced."

A second service member named Robbie Cordova took the stage. As an undergraduate at Fresno State, Cordova described himself as someone "without a lot of purpose." But a friend mentioned an internship that came with a scholarship, so he applied to the Climate Action Corps. He was hesitant because he didn't have any experience working in climate change-related fields, but he said the Climate Action Corps didn't care. They valued his passion. Now serving his third term, he says he keeps coming back because this job cares for him. "It cares about my mental and physical well-being, it cares about funding my education, so that I can explore any career path I want, and cares about turning me into a decisive leader," he said. The most important tool, he said, in fighting climate change is compassion. "Because we're not just taking on carbon dioxide," he said. "We're feeding the hungry, we're improving the lives of our fellow human beings and transforming our community."

At the event, the focus wasn't so much on the world each service member would be joining, but rather what they were about to get out of serving and how it extended beyond helping others. When Fryday took the stage, he echoed similar sentiments. "Because you stepped up to serve, you now have a bond with thousands and thousands of other service members across the state," Fryday said. "You will always belong to an incredible service family."

Together, fellows raised their right hands and repeated after Fryday the following words:

> As a California Service Corps member
> I am committed to make change in my community
> I will serve with humility, compassion, and integrity.
> Faced with apathy, I will take action.
> Faced with adversity, I will persevere.
> I will carry this commitment with me throughout my life

> I am part of a Service Corps
> Connected by a common mission: to build a California for all.
> I am a California Service Corps member
> And I will get things done.

In writing this book, I've thought long and hard about solutions. But a theme that has come up again and again is that the solution isn't an individual one. It's not what one person or sector can do. Creating systemic caring and making volunteering accessible for everyone is going to truly take a collective effort. The infrastructure for access won't only come from a cultural shift in mindset, but from the government actually building the infrastructure to make it happen—like the California Service Corps programs. But it could be that today's government officials, aside from those in Fryday's and Newsom's offices, are too set in their ways, putting community service and community engagement in the back seat. Instead, it's going to take a new generation to make a dramatic change happen.

9   *Why Mattering Matters*

Cat Moore would describe the first 24 years of her life as being caught in a state of chronic loneliness. Throughout her childhood and early adulthood, Moore found it difficult to connect with people. She struggled so much that in high school, as many of her peers were busy making memories with lifelong friends, she dropped out to homeschool herself in her bedroom.

Determined to live a less lonely life, when she went to college at the University of Southern California (USC), she decided to major in philosophy to understand more deeply the underpinnings of her feeling like she didn't belong. Perhaps not so surprisingly, it's here where she first got a taste of community, thanks to her late professor Dallas Willard, who made her feel like she belonged to a community of philosophy students. After graduating college, Moore says she was still pretty shy, and a tad bit lonely. A few years later, she got pregnant with her son, which is when she started going to coffee shops to just be around people, but not have to talk to people. But as any pregnant person knows, a pregnant stomach can easily become an unwanted icebreaker. Conversations with strangers ensued. Once her son was born, she continued going to coffee shops, which prompted more people to talk to her, as she had a young child in tow. Some would even line up in what she eventually called the "the latte window" as people waited for their coffee. "People would sit down, not knowing me at all, spill their life story, tell me whatever it was that was weighing on them," she told me in an interview. "Some would burst into tears. They would get up and say, 'Oh my gosh, thank you so much for listening to me.'" When she told her former professor what was happening, he responded, "Being with each other, as we are, where we are, is everything." Then he added, "We organize our lives to death to avoid it because it requires that we slow down and risk being known."

When a friend of hers at USC got word of what was happening with Moore at the café, they wanted to witness the action in person. Here was Moore, a person who suffered from severe loneliness all her life. Now, she had become a steward in her community simply by listening to strangers at a coffee shop. At the same time, her alma mater was dealing with a massive loneliness epidemic. Students were dropping out, having mental health crises, and overall, struggling to connect with each other. Sure, they could visit the campus counselor, but Moore said the problem was rooted in a general "relational culture." After seeing Moore's success, the university asked Moore if she could develop a curriculum based on what she had learned from the coffee shop and help create the conditions for friendship and community on USC's campus. "I didn't know if I could or not, but what I knew is that every single human is wired for connection," she told me. "We are built for this, and our lives don't flourish until we organize around each other."

This led Moore to create a curriculum to teach connection she now calls CLICK—an acronym for Connect, Listen, Investigate, Communicate kindness, and Keep in touch—and become the university's director of belonging, a position she still holds today.

Curious about the role that purpose and belonging play in our overall well-being and health, I traveled to USC's campus, one of the only college campuses I've found in my research that's making a tangible investment in a solution to the loneliness crisis.

Despite living in Los Angeles for a brief period of time in my mid-twenties, I've never visited USC. Excited to feel the timeless, buzzing energy of a college campus, I began my day by walking around before my meeting with Moore. Although the campus is situated in the middle of a city, there are plenty of places for solitude and there are places for gathering, like the campus village, which is lined with shops and restaurants. I stopped at a restaurant adorned with collegiate paraphernalia for a quick lunch. Signs promoting happy hour deals for football game days hung from the windows. Not only did I want a burger, but I was also hoping to make some casual conversation with a stranger, to learn more about my surroundings and maybe jot down a meaningful anecdote for my book. Instead, I was greeted by tablets sitting on top of every table. One person behind the bar said "hi" to me, but aside from that interaction, any opportunity to have a conversation with a stranger had been taken away by my app-based waiter. I sat and read before meeting with Moore, and I couldn't help but wonder, how many students have had to do the same before me?

Moore's office is appropriately located in the Religious Studies Department. As I walked up to the second floor to her office, I encountered a group of students wearing hijabs. Signs to promote interfaith activities hung from campus community boards. A group of students were eating a free vegan lunch that a faith-based organization provides weekly, with the goal of building community. If I had known about this program, I would have tried to have lunch here. I'm expecting a stuffy professor vibe from Moore in her office.

Instead, I'm met with something much cooler. She is vibrant, colorful, and kind. She has blonde hair with a tint of red, depending on how the light shines on it, and sky-blue eyes. She's wearing a T-shirt and jeans. Her bookshelf is filled with titles like *The Art of Gathering*, *How to Do Nothing*, and *Together*, the same ones that take up space on my own shelf. Sitting on her bright orange couch is a Barbie-pink inflatable flamingo, which she tells me we'll be using for an activity later that day. After our interview, we'll be venturing onto campus to do what Moore calls "random acts of belonging" to connect with students who might be feeling inescapably lonely.

As the director of belonging, Moore doesn't have an average day. Instead, she describes her job as "responsive," where she is focused on building relationships with people across the university to see what they're doing and how she can foster more belonging in those sectors. For example, political science professors might bring her in to discuss how connecting with others can help be the antidote to election anxiety. But many days, she finds herself talking to students one-on-one in her office, often promoting what students can learn in CLICK.

CLICK is based on a hybrid of Moore's intuition and scholarship. She knew the essential elements to create a sense of belonging were to listen to people, ask people questions, and do that regularly. But after Moore studied various disciplines of connecting, she made the first step in CLICK identifying people's "high connectivity" spaces. Where do they hang out already and spend their time? Moore says in seeking one's way out of loneliness, a person doesn't have to start from ground zero. The next step is to encourage people to find their "connection bridge" to other people, which can be as basic as looking at someone and saying "hello." After that, people should show interest and then communicate in kindness by caring for people they're hoping to develop a connection with in the future. However, it's the final step, keeping in touch, that is the hardest for nearly everyone, Moore says. She recommends creating a habit or "weekly rhythm" out of connecting with people. The general principle is once a week. But that's not always feasible for everyone. Ultimately, the goal is to create memories, connect as you are, and be vulnerable. "You have to be flexible, you have to be creative," she said, adding that it can sometimes feel like a lot of work. "But the alternative is things just fall apart."

When asked about the role kindness plays in CLICK, Moore said connecting is "simultaneously so simple and so complex." Listening, and asking questions, can be filed under social skills. But that's not all that goes into being part of a community and feeling like you belong. The special ingredient is caring for the other person. "Because no matter how many questions you ask someone, if someone doesn't actually care about you, you can feel that," she told me. In other words, where is your heart? Feeling a sense of belonging is a deeper, more spiritual practice, she said. "If you want people to see you, hear you, love you, and accept you, are you able to do that for other people?" she asked.

When it comes to belonging, caring for others is actually an empowering place to be, she added. "Your presence, nobody can take that away from you," she said. "You're a thermostat, not a thermometer, you're not just going in to take a temperature, but you can change the temperature." Moore is full of these quirky phrases.

Many students have social anxiety, Moore added. They are worried about looking awkward and being vulnerable. But if a person shows up knowing that their presence and their ability to care for others can make a difference, it allows the person to show up with strength. It doesn't have to unravel a person to their core when it's not reciprocated. The stakes are high when it comes to building a sense of belonging, Moore added, because our innate human desire to belong runs so deep.

While USC doesn't measure loneliness directly, the university does measure "belonging." The two are inversely related, Moore explained. When I ask how bad the situation is on campus, she said that one-third of students feel that they have a positive sense of belonging. When Moore attended USC, it was a competitive school. But today, it's "next level," she said, adding it has led to two outcomes: more transactional relationships among students or students being too busy with their extracurriculars to form meaningful friendships. What students need is more balance, Moore said, but the recognition of that alone requires a different level of maturity. Plus, if students aren't already immersed in a culture of balance, it can be hard to achieve as the only person pursuing it. "Who wants to be a loner, who's balanced, when everyone else around you is achieving?" she asked.

This kind of culture in our education system is bred at a young age. As the mom of a middle schooler, Moore sees it happening among her son's peers. Unfortunately, it usually takes a person reaching a point of crisis for them to understand that their existing connections aren't strong enough. "You realize you have a need for a deeper level of vulnerability or support, and then you realize you don't feel comfortable reaching out to the people you're connected with," she said. "And then you're like, oh, I don't really have that kind of friend."

When I asked if CLICK is helping students, she surprised me by revealing it's on hiatus. Not because it wasn't working, she said, but because the demand was too big. Considering Moore's hours, she could have only 15 to 30 students enrolled at a time. Currently, she's trying to figure out how to make it more accessible to students and more universities. Recently the Massachusetts Institute of Technology adapted it and is offering it to their students, which she is happy about. But at the moment, she is trying to shift her focus to what can be done immediately to solve USC's loneliness crisis. "I'm really into the small things that anyone can do and have immediate effects," she said. "People say we need systemic change, and we do, but I've seen how much life it brings people when they realize, oh, I can go take this handful of wildflower seeds, go sprinkle them in the cracks of the sidewalks, and literally change this whole environment." It's not an either-or situation, she said.

This is where the inflatable flamingo came in. Part of the problem of loneliness on college campuses, Moore said, are the high-stress, performative aspects of college life that make people feel like they can't mess up and play. She wants to detangle that culture with a playful activity. In CLICK, she uses flamingos to demonstrate that it's possible for people to find their fellow flamingos. "One of the reasons why people feel so lonely is because they don't know other people are lonely, too," Moore said. "Flamingos, they're weird animals and they live in the flamboyance of other flamingos." No matter how weird someone feels, there is always someone else around with a similar feeling.

Moore goes on to explain that we are about to head out and give away free lollipops as our first "random act of belonging." "It's string-free generosity," she said. "I just want to have one message out there, that says we aren't trying to convert you to anything or get you to buy something, we're just literally out here checking to see if you're finding friends and community." And also, who doesn't want a free lollipop?

It turns out, not many people. I returned to the USC village with Moore, a student volunteer, and another student who generously volunteered to wear the inflatable flamingo suit. Unsure where to go, we started the campaign in front of Target, where many students shop. The two students began by simply asking people, do you want a free lollipop? It was an offer many students rejected at first. Despite a slew of "no's," Moore reiterated that this is an experiment and it was okay if it didn't work out. But slowly, some people began to show interest and express curiosity: what is this for? they asked.

It almost immediately occurred to us that many people thought we were trying to sell them something, a realization that was followed by a lot of sadness. It meant that people weren't used to receiving something for free out of the kindness of a stranger's heart. It meant that people were stuck in a pattern of transactional relationships with strangers, as Moore suspected. But not everyone rejected the free lollipops. One woman accepted a lollipop and asked why a student was wearing an inflatable flamingo. The other student told her that they were conducting random acts of belonging and wanted to let people know that they belong by doing something nice for them. And if you were struggling to belong, Moore was available to listen to you. The woman was very moved and told the friend she was waiting for, who happened to have two cats, about the random acts of belonging. It ended up that the cats drew a lot of attention and more people became curious. What was a difficult start turned into more people saying "yes" and accepting the free lollipops. Not only did people feel cared for by receiving them, but also, it truly was an effective way for Moore to check in with some students on campus. It made them feel safe. For example, when she asked a group of foreign exchange students if they felt like they "belonged," they revealed that it was really difficult to make new friends and adjust to life at USC. "We were just talking about how hard it is," they said.

Once the lollipops melted in people's mouths, and hearts, Moore asked the students to hand out yellow flowers. Admittedly, these were more of a hit than the lollipops. Many students seemed to be truly taken aback by being gifted with a free flower. I also participated in handing them out. True to my research, it felt good to see the smiles on the students' faces. For those who would stop and listen, Moore once again found opportunities to connect with students who were struggling with loneliness. One student told Moore, as tears welled in her eyes, that the free flower made her day. She said she had been struggling, that a lot was going on in her life. She looked Moore right in the eye and said, "This really helps," with such sincerity. "Nobody might have engaged them if we weren't here," Moore later said about the interaction. By making a difference to one person, and showing them that their presence mattered and they belonged to this community, our work here was done.

• • •

As a granddaughter of a Holocaust survivor, I've long been interested in the work of writer and survivor Viktor Frankl. Logotherapy, which he founded, can be summed up like this: a sense of meaning and purpose in life is the central force for having good mental health and living a healthy life. In his book *Man's Search for Meaning*, Frankl warned that with society's movement toward automation, more people would experience boredom.[1] Of course, this isn't always a bad thing. Boredom can lead to creativity. These days, it's almost a privilege to be bored. But it becomes unhealthy when people get sucked into what Frankl called an "existential vacuum," where they lose a sense of purpose in life. Fortunately, Frankl had an idea about a way out of this, a way that people could find their place in this world and have such a deep sense of meaning it could be felt by their simply existing. "The more one forgets himself by giving himself to a cause to serve or another person to love, the more human he is and the more he actualizes himself," Frankl wrote. Purpose, Frankl wrote, isn't something that can be discovered by getting lost in one's own psyche. Instead, it's something that can be found by connecting with and caring about other people.

Decades later, Frankl's theory holds up. A book published in 2007 titled *In the Course of a Lifetime* analyzed the IHD Longitudinal Study, which followed 200 people in Northern California every decade of their lives, starting in adolescence.[2] It was the first study of its kind to examine the lives of humans over a long period of time to figure out how people live and what matters when it comes to longevity. While the authors analyzed the role religion played in people's everyday lives, they found that those who showed a sense of purpose early in life, contributing to the well-being of others, were healthier and happier later in life. "We found post-mid-life that while church-based engagement certainly nudges generative activity, our spiritual seekers also found their own ways of purposefully helping others," Michele Dillon told me via email in 2017. "There are different paths to helping others and different ways of doing so; though in later life the 'religious' were happier than the 'spiritual'—but this seemed in part due to a different way of construing 'happiness' (i.e. it's not necessarily the same as personal fulfillment) and different ways of coping with personal stress."

What's at stake, to our individual and societal health, if we don't feel a sense of purpose? Can we actually get sick if we feel lonely and as if we don't belong? To find answers, I hopped on the 405 freeway, endured bumper-to-bumper traffic, and meandered across Los Angeles to the University of California campus.

In June 1981, the U.S. Centers for Disease Control and Prevention published an article in its *Morbidity and Mortality Weekly Report* titled "*Pneumocystis* Pneumonia—Los Angeles," describing five healthy gay men in Los Angeles who had rare lung infections, in addition to other infections indicating their immune systems weren't functioning properly.[3] By the time the report was published, two of the five men included in the report had died. Soon after, the rest of them would die, too. This eventually marked the first official reporting of what would later become known as the AIDS epidemic.

Early in his career, scientist Steve Cole took an interest in the AIDS epidemic. He worked on a nine-year longitudinal study following 80 HIV-positive gay men. Every six months, the men in the study gave blood, sat for interviews, and filled out questionnaires. Every six months, fewer and fewer of them showed up. Cole wanted to understand why the disease killed different people at different rates. Could it be sleep quality, exercise habits, socioeconomic status, or a history of anxiety and depression? None of these seemed to yield significant results. But then Cole and his colleagues had an idea: compare openly gay men to those who were hiding their sexual identity. They found that closeted gay men died faster than gay men who were open about their sexual identities. This research, and more that followed, led Cole to an unprecedented understanding of how viral genomes worked as a system and hinted at the devastating health consequences of feeling lonely and like you don't belong.

Decades later, at a think tank meeting in Chicago, Cole was talking about his work in viral genomics when a man came up to him and asked, "Can you do this for the human genome?" In the back of his mind, Cole thought "no way." The human genome is 20,000 genes—but he lied to him because the man who asked him was John Cacioppo, the author of *Loneliness* and one of the founders of the field of social neuroscience, who had already identified loneliness as a risk factor for poor health. At the time, loneliness was a niche topic. Cole wanted to work with him, and Cacioppo already had blood samples from lonely people. Cole went ahead and extracted a leukocyte RNA sample, and to his surprise he was able to note differences in the RNA profiles of lonely and nonlonely people almost immediately.[4]

In the RNA samples of lonely people, the genes involved with inflammation were more active. The genes involved in anti-inflammatory responses—the genes that served as major determinants for how far HIV effectively spread—were suppressed. The RNA profiles of lonely people also lacked a healthy number of interferons, whose job is to stop viruses from replicating. These observations helped lead Cole to a well-known narrative we know about lonely people today: lonely people are at an increased risk for major chronic diseases. At the time, researchers were just beginning to understand how inflammation was, what Cole called, a "generic fertilizer" for the production of many diseases, ranging from metastatic breast cancer to Alzheimer's. "It was a remarkably clear portrait," he told me in an interview on UCLA's campus.

Cole's first published study on the genomic profile of a lonely person received a positive response not only from the science community, but also from the general public. He said he received many letters from people sharing stories about family members and friends who had become shut-ins after a betrayal or how something led them to chronically lose faith in humanity, and then they contracted a disease. To have people say that they cared about his work and encourage him to continue his study kept him going.

He used the same method to test other social adversities, such as stress and PTSD, and proceeded to measure inflammatory gene expression and less antiviral gene expression. Again and again he noticed a pattern of more inflammatory gene expression and less antiviral gene expression, regardless of the type of adversity. This launched another decade of work to understand what was happening neurobiologically, "which is when people feel threatened or insecure, they don't have to think about it, their brain stem just turns it on and activates fight or flight biology," he said. One of the places that's impacted the most is the bone marrow, which is where our immune cells live and are constantly regenerating. However, in periods of chronic threat or uncertainty, the bone marrow ramps up the production of these cells, which are called monocytes. The consequence of overactive monocytes is that when a real threat enters the picture, like a concussion or tumor, they respond with extra inflammation, leaving the body more vulnerable to a potentially worse progression of the condition or disease. A classic example is cancer, Cole explained. Usually, cancer starts by wear and tear to the DNA. That is usually not enough to kill someone, but it's enough to form a tumor. If the tumor metastasizes, monocytes can act as catalysts to assist the cancer cells in moving around and successfully spreading to other parts of the body. The overactive monocytes could be the result of living in a state of chronic loneliness. To be clear, in the decades of Cole's work, he hasn't found that loneliness is the cause of disease and death. But what he has found is that it can help create the environment to make a disease worse.

Cole was kind enough to show me his lab at UCLA. The first rule was not to touch anything with my bare hands. The second rule: don't touch anything with my bare hands. Chemicals were everywhere, and I wasn't wearing the right personal protective equipment. Prior to our interview, Cole warned me that the lab was going to be "boring." Sure, it lacked sci fi-esque decor, but it was far from making me yawn. The sounds of machines buzzing filled the air. Computers and piles of papers fastened with binder clips were stacked on top of desks and shelves. I felt as if I was surrounded by a constant sense of discovery. A pair of scientists in white lab coats handling blood samples in tiny test tubes sat in front of a well-lit desk. These days, Cole said, he and his colleagues analyze blood samples with the help of their robot friends, a change from when he first started his career. While it may appear to be less exciting, it does allow them to do more work and take on bigger projects, and it also saves them from the grueling task of having to analyze blood samples by hand. Each robot in the lab is named after the last human who had to do that task before a robot helped automate it.

One robot had plastic tubes mounted en masse onto its body. Despite my anthropomorphizing description above, there is nothing humanlike about this robot. It looks and sounds more or less like a copy machine. Inside its gray boxes, the robot separates the blood samples and takes out particular molecules that the scientists program it to separate, like a specific protein or RNA. Depending on its nature, a small study might require the analysis of 20 blood samples. A big one could include thousands.

When I asked what they were working on, Cole explained that the scientists were studying blood samples for a business school professor in Japan who was curious about how a change in workplace culture was affecting staff well-being. Historically, a Japanese company took care of each employee as if they were a family member. Employees would spend their whole careers at one company. Getting fired was rare, unless the company was going out of business. But now Japan was transitioning to a U.S.-type at-will employment system. While it's incredibly productive economically, Cole said, it's a constant stressor for employees. And this business professor was keen to know how such stress under precarious circumstances affected gene expression. He didn't have the answers yet, but one could easily see how this situation fit into the general theme of Cole's work: living in a chronic state of stress increases inflammation in our bodies, which acts as fertilizer for severe disease. We have built a world that is highly economically productive, he said, but it's not good for human health. It leaves people living in a "chronic, low-grade drizzle of stress biology, day in and day out, month after month, and over years that accumulates into a heart attack, metastatic cancer, or a faster onset of Alzheimer's," he said.

The work that has been conducted in Cole's lab has led to unprecedented findings about loneliness and human health. For example, during the COVID-19 pandemic, Cole and his colleagues were curious to know how social isolation could be affecting antiviral immunity—despite the intent of sheltering in place to protect people from the coronavirus.[5] At the time, perceived isolation—in the form of loneliness—was known to weaken the immune response. But what about actual isolation? To find the answer, researchers relocated 21 adult male rhesus macaques from their communities to an isolated cage for two weeks, similar to how many humans went from living their everyday social lives to living in an isolated apartment. Sheltering in place for the monkeys was associated with a 30 to 50 percent reduction in the circulating immune cell population, which showed up in their blood samples as quickly as 48 hours after the start of isolation and persisted for two weeks. Immunity didn't resolve until four weeks later, after they had returned to their respective monkey communities. Notably, when some of the monkeys were given the chance to care for a younger monkey in isolation, their immune responses were more robust than those who were in complete isolation.

Of course, these were monkeys and not humans. Conclusions that can be drawn aren't absolutes. Plus, for humans, caregiving can be stressful, Cole said, but that's because people are forced to do it on their own. "It's tremendously taxing on your physical and attentional resources," he said. A sustainable version of caregiving is where people are allowed to rest from it and get some rejuvenation. It doesn't mean that caregiving doesn't have any value or aspirational component. Caregiving can lead to humans utilizing what he calls the "virtuous reflex." "The trick is really, how do we restructure our everyday life in a way that supports that kind of caregiving collaborative community activity?" he asked.

The connection between loneliness, other forms of social adversity, and molecular gene expression is the fight-or-flight stress response from the body's autonomic nervous system. A human's fight-or-flight response, Cole said, was meant to operate only a small percentage of the time. In hunter-gatherer days, humans weren't stressed and worried all the time, like many of us are today. They actually spent an estimated one-third of their day communicating and learning from each other—that is, socializing. Eight hours of recreation a day, Cole said, is the lifestyle our bodies evolved to support. In that context, hunters and gatherers utilized their fight-or-flight response only while hunting, defending against a raid, or something similar.

Cole said he believes a lot of people feel alone because they don't feel cared for and they don't feel like they have enough people to connect with who care about the same values as they do. He said people don't have to necessarily volunteer with a group on a regular basis to find a community with shared values, but people can create these communities around art, nature protection, conservation, and scientific discoveries.

Historically, ecosystem-level threats—like natural disasters or war—have been what brought people together. But the longer America goes without any fundamental existential threat, Cole said, the more society will see "dissociation and atomization." As a result, people are glued to their screens in an effort to soothe themselves, which is good for our existing infrastructure in economics and politics, but not good for the nervous system. "It's an incredibly corrosive effect on the basic sense of safety, which is fundamentally based on trust in other human beings," Cole said.

When asked what the ideal molecular profile would be for humans today, he said it would be one with a low level of inflammatory gene expression, a more relaxed production of monocytes, and a higher level of activity of interferons.

How can this be achieved? Cole said it's not by intermittently relaxing or meditating, but by truly creating a life that's centered around purpose and connection. Purpose and connection translate to certainty to the nervous system. Purpose and connection act as a brake to the autonomic nervous system from firing off too much of a stress response. When a person is focused on doing something important for the world in general that they regard as important and that other people see as important, that brings that person into a community of mutual benefit, Cole said. "And that is the chief source of safety for human beings," he said. "The idea that all I care about is me and my well-being is an inadequate response, especially to the nervous system."

Cole said this all goes back to the evolution of the brain, particularly the ventral striatum. As we know, it was wired to feel rewards for food, but somewhere along the line it became capable of "hooking itself up" to other values, like doing something good for another person. "It turns out it's a trick. If you build an organism that is biased to do good for other members of a species, it will self-assemble communities," he said. "And those communities can do amazing things."

Like neuroplasticity, can the genomes experience plasticity, too? In other words, can people reverse the increased inflammation or low-level antiviral responses if they make the shift from loneliness to feeling safe with others by way of purpose and connection? Yes, Cole said. But it's tough in our culture. When it comes to potential damage, he warned, "It's not so much how quickly social adversity gets into the system. It's how long it stays there." Without purpose and connection, people may feel like they are just a "molecule floating around." This isn't living a thriving, happy existence, Cole said. "This is just waiting to die."

A way for a person to ensure that their genes are working in their favor, Cole says, is to make sustainable life changes, to become connected to a sense of purpose and a community that cares about them. Cole said this is one of the worst aspects of the downfall of religion in America. Church and temple goers felt that they had their places of worship to support them. Through a default of trust and faith, fewer people are now active participants in religion.

Prioritizing caring and a more virtuous lifestyle in our culture will set up future generations for success, Cole said. But it won't happen through a literal passing down of gene expression. Since only two strands of DNA are passed from parents to their children, it might be a tough process to depend on. However, by our creating infrastructures that aren't built on scarcity and supporting a more regenerative and reciprocal lifestyle, future generations will be positioned to have better health.

Knowing everything he does, I asked Cole, how do you live your life differently? He said that many findings that materialize out of research are "things your grandmother would tell you." "We keep discovering that our grandmothers were right," he said. "If you want to understand disease and death, don't track the number of minutes you're sad or happy or guilty or something like that. Track the number of minutes that you're lonely."

As I reflected on my meeting with Cole, it became clear to me that our individual health is affected by societal health and our external environments. The rise in disease in the U.S. could be attributed to the rise in loneliness and an overall lack of feeling safe in society. As the science has shown throughout this part of the book, helping others and being kind is good for our health. It can be the pathway out of loneliness. It's a conduit to feeling safe and forming the deeper connections that we crave. But just as Cole noted that our grandmothers were right, I can't help but think that the culture our grandmothers grew up in was very different from the one we live in today. If we can't create sustainable change in our culture, where kindness and caring can thrive, then individual approaches will only act as Band-Aids. While it's nice to see interventions happening at a university level, like I saw with Moore at USC, I can't help but think this needs to be happening in early education. Perhaps it's the same with changing culture, too, as Joan Morgenstern alluded to at the beginning of this part (chapter 5) regarding building habits of kindness from an early age.

III *Sustaining Systemic Caring*

10 *The Power of Empathy*

Formerly a high school teacher, Glenn Manning saw firsthand the intense pressure young people face to achieve academically. There's no question, he told me in an interview, that there is a connection between mental health struggles among today's youth and an intense focus on academic achievement in our culture. In our educational system, there's a message from society to focus on one's own happiness and personal satisfaction, even at the cost of not caring for others. "It's in the way that students hear us speak," he explained. "And it's in the incentives that our systems put in front of young people that they respond to and internalize." When the focus in education is solely on academic achievement, the system unintentionally sends the message that other values, like altruism, aren't important.

The issue could be manifesting in unprecedented ways, like the current absenteeism crisis. Today, high school students are chronically not showing up at school. Unlike when fictional high school senior Ferris Bueller pretended to be sick in order to play hooky and persuaded his friends to join him on a tour of Chicago for one day, today's epidemic is much more sinister. According to data from the U.S. Department of Education released in 2023, during the

2021–2022 school year, an estimated one of five students missed school for a total of almost four weeks.[1]

Manning told me he doesn't want to pinpoint one reason behind the crisis, like mental health issues among youth or the pandemic. But what he does see is that when high schoolers are in an environment in which they feel safe and cared for, they are more likely to show up at school. When people feel safe, and make others feel as if they belong, values like altruism thrive. Manning said it's not good enough to leave such an environment to chance. When we do, he said, issues like hate, bigotry, harassment, and perhaps even absenteeism arise. "We built a moat and diminished everyone's capacity for caring and for seeking and understanding justice," he said.

Notably, Manning identified an interesting trend among families in communities where academic achievement is highly valued. In his research with Harvard University's Making Caring Common project, when parents are asked what they value the most for their kids, they will say caring. But when asked about what other families value the most, they will say academic achievement. "We're ascribing values to other people in the school community that they don't actually hold," he explained. "And so a gap widens, and that's how we get this real sense of competition between family groups when no competition needs to exist."

Psychologist Richard Weissbourd, a senior lecturer at the Harvard Graduate School of Education, directs the Making Caring Common project, which was founded on the idea that in today's parenting and education culture, individual success takes precedence over caring for others. "If you look at all the problems in the country right now, they have many sources," he told me in an interview. "And one source is that we haven't prioritized caring for other people and caring for the common good."

It's notable that people at Harvard University are trying to elevate values like altruism to the same level as academic achievement. Harvard is one of the hardest universities in the world to get into, and its students are no strangers to ruthless academic achievements. The average high school grade point average (GPA) of admitted students at Harvard is around 4.2. But in their program, Manning, Weissbourd, and their colleagues have developed tools and training for K-12 students and teachers to cultivate a culture of caring—not rigorous academic achievement.

One example of an activity that the program developed for schools is called relationship mapping.[2] In this activity, the Caring School Network helps identify students who might be at risk for any personal or school-related issues. It does this by asking every teacher, counselor, and front office staff member to identify the relationship they have with every student at school. They do this by simply putting a star next to a student's name. Some students have numerous stars, while others have zero. Weissbourd said the goal is for every kid to have a trusting, caring relationship with an adult at school. "Part of the purpose of this is just to be intentional and systematic about it," Weissbourd told me. "Everybody sort of believes in relationships, but the relationship mapping strategy provides a map ensuring every kid is anchored to at least one adult." Once they've collected the data, they can see if there is a student who doesn't have a relationship with a well-trusted adult. From there, they can make suggestions about who can start to develop one. Once adults have volunteered to develop a relationship with a student, they meet regularly with the student throughout the year. The team has taken this to another level by expanding relationship mapping to peer-to-peer friends, too. "It's not just having somebody care for you at school, it's you caring for other people," Weissbourd emphasized. "Focusing on other people is healing."

Not only could this help lead to a safer environment, but like Manning said, it could help ease the absenteeism crisis. Weissbourd said there are compelling reasons why many students might be missing school. Maybe they have to help care for a sibling, due to America's childcare crisis, or help a sick relative. But certainly relationship mapping could help bring back those students who feel like they don't belong.

Another example of an activity they provide to schools is deep listening.[3] Making Caring Common suggests this specific activity for kids in grades 6 through 12. In their lesson, students engage in a series of exercises with different conversation starters. During this time, they practice "being active, authentic listeners with a partner." The goal is "to make the speaker feel heard without reciprocating in the conversation." Speakers are also encouraged to become more comfortable with sharing about themselves and expressing their feelings.

Each lesson isn't long—only 15 minutes a week for five weeks. The first lesson is simply a brainstorm session, where the teacher might start out saying to students (quoting here from the Making Caring Common materials):

> This week we are going to begin a series of lessons focused on our strengths as listeners. I hope that each of us could name a time when someone really listened to us and the powerful impact we felt knowing that they were listening. But I'm sure we each could also name a time when we felt like someone really wasn't listening to us and the way that made us feel. It can be especially hard to communicate effectively in a virtual setting, which makes being

aware of good listening practices even more important. Today we are going to begin by thinking about what makes someone a strong listener.

Students can proceed to journal about what a good listener looks like to them and then share what they wrote with the class. The teacher can then identify good and bad examples of being a good listener. In the same lesson, the teacher is encouraged to guide the students to better understand the role of empathy in listening by noting that it can make "a big difference to a speaker if they are feeling affirmation from a listener." As the lesson plan states, "That's empathy: when we understand the thoughts and feelings of others and show that we care."

In the next four lessons, students work in pairs. The teacher proceeds to tell students that the first speaker has 1.5 to 2 minutes to talk, with no contributions from their partner except practicing the listening skills they learned during the first lesson. "For the listening partners, you'll remind them of the three listening skills, and remind them to 'think and feel and listen' for the nuances in their partner's responses," the curriculum states. Teachers are encouraged to acknowledge it's a little awkward, and they are provided with prompts that aren't so serious, such as "What did you do this weekend? What's your favorite memory or vacation so far? What are you afraid of and how can you conquer your fears?"

Manning said the idea is to be open to others' ideas and hear them, "as opposed to merely reacting to them in order to understand their perspective."

Listening, Weissbourd told me, is a fundamental human skill. "It's at the heart of empathy, is the art of reciprocal mutual good relationships," he said. It's also the antidote to the hyperindividualism in the U.S. right now. When I asked if listening is also a form of caring, he said, "Of course."

. . .

In June 2023, I was invited to attend an event called Connections at the Capitol in Washington, DC. Unfortunately, I couldn't attend in person, but I was able to join virtually. I was interested in the event because it was yet another attempt by the federal government to address loneliness through the lens of it being a public health issue.

As I discussed in part one, the government has taken a stance on reducing smoking and other public health threats, but what is truly being done to address loneliness and the lack of social connectedness, meaning, and purpose that Americans feel today? Unfortunately, today's politicians don't exactly set good examples of how to work together and be kind to one another. In fact, these days they give us more of a master class on how *not* to treat each other. But at the event I was heartened by the speech by Connecticut senator Chris Murphy, who made the connection between loneliness and political divisiveness.[4]

"There's a real political cost to ignoring this epidemic of loneliness," Murphy told the crowd, arguing that the anger many people feel, as a result of feeling lonely, allows "demagogues" to prey on people who are vulnerable. "I just don't think it's a coincidence that you've seen a rise in fringe politics in this country, a rise in conspiracy theories, a rise in political violence, as we have seen an increasing number of Americans reporting that they are feeling lonely," he said.

I don't either. Murphy said part of the strategy for tackling loneliness should be to give people more free time to "stretch their legs" and find community. "Paying people more, allowing them to have one job, allowing them to leave work at five," he said. In other words, putting people's needs and humanity first, not the needs of companies and their profits.

Then he acknowledged the current state of politics. How it's also become more angry, more "shouty," he described. And more polarized. Indeed, a study in 2020 found that America is "polarizing faster" than other democracies.[5] Murphy said he thinks that politicians and Americans alike need to take a step back from talking about specific policies and instead talk about how they feel. "If you have a conversation about the things you feel, we will find out pretty quickly that people on the right and the left are feeling a lot of the same things," he posited. "They are feeling run over by technology, they are feeling disconnected." There are a lot of lonely people who want former president Trump to go to jail, Murphy said. And there are lonely people who want him to be president.

After hearing Murphy speak, I couldn't help but think that our politicians would greatly benefit from one of Making Caring Common's deep listening activities. At least, by starting at a young age in a handful of our schools, we can have hope that we are beginning to change the tide on how people talk and listen to each other, despite their differences. Because if the only example for kids today is watching how listening is (not) happening at the political level, they won't have a chance at learning how to be good listeners, which is the first step to building empathy and inspiring us to be kinder.

• • •

Scientists believe the emotion of empathy can trigger altruistic behavior. As the brains behind the Making Caring Common project said, listening is the first step to empathy. If we listen to each other, we will build empathy. If we experience empathy, we will become more altruistic. The problem is that, as many headlines have noted over the past decade, the country is in the middle of an empathy crisis. In 2011, researchers published a study that found college students were about 40 percent lower in empathy than their counterparts of 20 or 30 years ago.[6]

"We found the biggest drop in empathy after the year 2000," said Sara Konrath, a researcher at the University of Michigan Institute for Social Research at the time. Researchers suspected that social media might be to blame, as technology was leading to more disconnect in our culture. "The ease of having 'friends' online might make people more likely to just tune out when they don't feel like responding to others," one researcher said at the time.

But if technology is to blame, is it the education system's responsibility to fix it?

Manning noted the lack of resources and staffing shortages in America's schools. How effective is it to go to teachers and ask them to learn and teach a new lesson plan on caring, especially when they're under so much pressure to have their students achieve academically? Manning said he believes it is the education system's responsibility to teach caring and pointed to the founders of America and many great educators, who were clear that both intellect and character were a central part of the American education system. If we teach kindness and caring correctly, Manning said, America can have a thriving democracy. "I think we see what it means to promote one and demote the other, we see the divisions in our society, we see the fractures, we feel them," he said. "So this is a return to a historical mission."

Manning added that schools are one of the only remaining places where people can experience a sense of shared reality, especially in the face of individualized technology. And as religion continues to decline in America, with fewer people going to church, schools are one of the last places where people learn cultural norms, which is precisely why they could be the best place for systemic caring to begin.

After talking to Manning and Weissbourd, I thought back to Stephen Porges's theory on how our nervous systems are constantly in a state of fight or flight because we live in a culture that's constantly evaluating us. How this keeps us in a survival of the fittest mindset. How Manning said each family wants to value caring for others over personal success, but believes other families are prioritizing success over caring. How the first steps to systemic caring are listening and building empathy. And how America's gun crisis adds another complex layer to the idea of feeling safe in school. But let's say that for this thought experiment, this could work. That by learning to listen, we could build more empathy and, as a result, become more altruistic as a culture. That we could experience a durable state of bounded solidarity without it being based on a group being disenfranchised or in crisis. What's the neurobiology behind that, and is there something to learn from the science of feeling safe, as well?

11 *Nature's Fire Extinguisher*

A quick Google search states that oxytocin is the love hormone, but it's much more than that. Oxytocin is a neuropeptide molecule that is central to the biology of feeling safe, and this feeling starts with the mammalian mother-infant bond. Like my own daughter, baby mice cry to get their mother's attention. Studies have shown that maternal mice are quick to respond to their infants; mother mice even respond to baby mice that aren't their biological kin. Neuroscientist Robert Froemke led a study to better understand why mice that never had any offspring never seemed to respond. Based on previous research suggesting oxytocin was a key part of bonding, Froemke and his colleagues injected oxytocin into the left auditory cortex of these mice, which in the human brain is responsible for recognizing social signals. The virgin female mice responded to the crying baby mice sooner than the mice that were given only a shot of saltwater.[1]

This study suggests that oxytocin is a hormone that enables us to care. It also acts as a moderator of the autonomic nervous system. Dr. Sue Carter, a distinguished university scientist and Rudy Professor Emerita of Biology, at Indiana University, is a pioneer in understanding how oxytocin functions. One of her most significant breakthroughs came from a collaboration with zoologist Lowell Getz, where they documented the physiological underpinnings of social monogamy in prairie voles. For years, Getz was convinced that prairie voles were living as monogamous pairs. Carter noticed that in the world of prairie voles there were other factors at play, like the lack of an ovarian cycle, which suggested that reproductive hormones weren't the only hormones contributing to their monogamous culture. Carter's research led to the discovery that oxytocin levels rise when a prairie vole meets another special prairie vole, suggesting that their monogamy is not based solely on the necessity for sexual reproduction. It has to do with feeling safe.

Carter emphasized to me in an interview that, true to its role in social bonding effect, oxytocin does not act alone. The hormone needs a partner, and its partner's name is vasopressin. Both work together to regulate the human nervous system, a highly evolved system that isn't yet quite fully understood, Carter said.

What we do know is that vasopressin is the more primitive of the two hormones and has been associated with the neurobiology of fear and anxiety. Vasopressin is a stress hormone, a hormone that promotes inflammation. Oxytocin has the opposite effect. Both evolved from a common ancestral peptide. "Remember, we are based on this oxygen story," Carter told me. "We live in an environment that makes us vulnerable to oxidative stress, to rusting, but we have a defense, and my argument is that oxytocin is nature's fire extinguisher."

I couldn't stop thinking about Carter's referring to oxytocin as nature's fire extinguisher and about the way oxytocin works in conjunction with vasopressin. The two need each other, and in a way, they need to care for one another. Even at the molecular level in the human body, it's a constant delicate dance, a push and pull, between a state of aggression and calm. But when the two work together in the most optimal way, their default is a state of calm. And when that happens, humans can do amazing things—all because they feel safe and cared for by each other.

"My personal, private theory is that oxytocin is there both to help us make social connections, which is going to help everything, but also to protect us against overreacting to vasopressin," Carter said. "Oxytocin used to be thought of as a female hormone, but what you have is a hormone that sits at the center of resilience and survival."

In other words, what is thought of as being feminine—that is, being caring and nurturing—is actually the biological underpinning of strength. In search of a microcosm rooted in a caring-first mindset, a restorative state, I circled back to the social prescribing movement.

・ ・ ・

When studying to be a psychologist, Dr. Elizabeth Markle did one of her many trainings at Cambridge Health Alliance, part of Harvard Medical School. Working in the emergency room, Dr. Markle saw patients and also provided mental health support for physicians and new residents. Time and time again, she saw that therapists were prescribing what she described as "behavioral prescriptions," like exercise, better nutrition, or therapy, for mental health support. But aside from delivering these "prescriptions," the therapist had no way to follow through and see if they were being followed. Also, the ability of a person to execute a prescription often depended on their privilege. To Dr. Markle, it felt like she was frequently performing "health care theater."

"It just became so clear to me that this was a prescription with no pharmacy, right?" she told me. This was, in Dr. Markle's words, a "delivery problem and an equity problem"—an equity problem because the patients who could hire their own health coach, take Pilates, or eat organic food had a way forward because they had money. "The patients who didn't, they didn't have a way forward," she said.

That's when she started really thinking about what it would look like to create this so-called equitable and accessible behavioral pharmacy. What would be a way to effectively deliver these behavioral prescriptions? It wasn't until eight years later that she was able to pilot her nonprofit called Open Source Wellness, located in Oakland, California. Since then, there have been various iterations, she told me, but at the core of Open Source Wellness are four pillars: move, nourish, connect, and be. Human connection, Dr. Markle said, is one of the most important parts of each session.

Today, the nonprofit runs weekly groups that anyone can join. While it costs $125 a month for a membership, many people who join are given a prescription to join a group for three to four months from their primary care physician at a low-income medical center.

Notably, the group is transdiagnostic, meaning it's not specifically for people with depression or diabetes or hypertension. "It's a group for people who, in their humanity, have strengths and challenges with their well-being," Dr. Markle told me. "We have folks whose primary challenges are related to substance use, whose primary challenge might be primarily physical complaints. We have plenty of folks whose primary challenge is mental or behavioral health, and the majority have multiple chronic conditions."

Dr. Markle was kind enough to allow me to see a group in action, virtually. On a Tuesday night, I logged on to Zoom. A high-energy

Open Source Wellness coach welcomed me. Music played in the background as she greeted each client, one by one, by name. I instantly took note of how nice it felt to be welcomed by name, even though I had never attended a group before—and I was attending as a journalist. I've been to countless online and offline exercise and wellness-centered classes in my life, and rarely have I witnessed everyone be welcomed by name. Rarely have I felt cared for in a similar virtual setting. Online clients varied in race, age, and sex. There were about 20 of us in total. Once everyone logged on, the coach provided an overview of what to expect over the next hour and 40 minutes. She also emphasized this wasn't a place where we were going to just talk about things, but actually do them.

First, there was an icebreaker. The question was "What's your jam?" Similar to the subjectiveness of wellness, some people answered strawberry, raspberry, or orange marmalade. Others named their favorite songs. I said any song by Fleetwood Mac, my favorite band.

Next up was the move portion of the program. Here, everyone was asked to participate either from their seats or standing up. The coach emphasized we should do whatever we felt comfortable with. After all, the Golden Rule at Open Source Wellness is "take care of you." Everyone's cameras were on, by the way, but there definitely didn't feel like there was a need to impress. As the music started, a fitness instructor guided us in an aerobics routine. Side step, front step, and a few grapevines later, each participant was asked to do their own dance move when their name was called. I'm not going to lie, I felt nervous and ended up doing my own personalized version of a hand-rolling wave. I felt silly, but never embarrassed, perhaps because the space inherently felt safe, and as a result, oxytocin was keeping the vasopressin in my body in check.

The dance portion lasted about 25 minutes and I certainly felt a shift in my mood. I felt less self-conscious than when I arrived to be virtually in front of a group of strangers. I also felt energized. Next, a new coach led us in the "be" part of the program. Here, we were told to breathe in kindness and breathe out whatever didn't serve us. Then the coach opened the floor and asked, "What is self-care?" People answered by providing their own definitions, which usually centered around taking care of oneself in very basic ways. One person said making sure she did her ADLs—an abbreviation for "activities of daily living"—like hygiene, walking, and eating, every day. The coach chimed in and praised everyone for participating, saying that self-care can be taken in any direction.

"Self-care are actions to take to preserve or improve well-being," she said, emphasizing that it's something to practice to prepare for when a stressful situation might happen. She furthered her point by moving on to a journal exercise, and we were asked to write down what fills our proverbial cups. Some people said focusing on their spiritual practice, like breathing exercises, while others said some form of movement or listening to music. Then the coach asked, "What empties your cup?" Worrying or feeling stressed, one woman answered. The coach went on to say that the goal of self-care is to fill up one's cup so that what empties one's cup doesn't completely empty it. She held up a cup of marbles to illustrate her point. "If this cup is full," she said, "and I take one marble out, it's still full. But if it's barely full, and one marble is taken away, then the cup is empty." I loved how the idea of self-care being promoted here wasn't a way to feel good, happy, or pretty, but instead a way to build and sustain resilience.

The third part of the program was the "connect" portion. After all, the tagline for Open Source Wellness is "Community is

Medicine." In this section, people broke out into smaller groups, whom they've been meeting with on a weekly basis with a health coach. In these groups, they discuss their progress with their goals, which can be as simple as walking a couple of times a week for, say, someone who is struggling with diabetes. This is a time for people to get personal. Understandably, I wasn't invited to a breakout group. But I did reconnect with Dr. Markle to talk about what I had just experienced.

Curiously, I asked about the roles altruism and kindness played in the program. Instead of being a main fixture, they still had a presence, but I couldn't quite put my finger on how. I felt it, but they weren't explicitly talked about. Dr. Markle excitedly said she needed to send me a photo that had three words on it: generosity, abundance, and contribution. In her working theory, Dr. Markle believes that if health coaches meet people with generosity, through their own vitality and vulnerability, participants will start to feel a sense of abundance. "Our coaches just love people, they love them some more and then they love them deeper, and that creates this experience of abundance, like there's enough to go around and you don't have to compete with the other people in your group," she elaborated. "And then, the participants very naturally want to be a contribution to others." This certainly tied back to the *listen, build empathy,* and then *become* altruistic model I uncovered at Harvard.

Dr. Markle added that at the end of each session the coaches open the floor for "offers and requests." Someone might say, "I have some zucchini in my garden—who wants one?" Another person might say, "I'm going to try to get up at 7 a.m. every day and go for a walk, and I want an accountability partner." "We're trying to grease the wheels for that kind of generosity and abundance and the opportunity to know oneself as a contribution," which can be quite powerful, Dr. Markle said.

Indeed, Dr. Markle nailed exactly the secret sauce I was sensing: abundance. At face value, Dr. Markle's curriculum could be packaged to attract and be sold to more affluent wellness types in the San Francisco Bay Area. But without a foundational focus on kindness and generosity, it wouldn't have the special touch that I experienced that evening. Like I said, from the very beginning I felt safe. I felt an undercurrent that people there really cared for each other, even though technically the participants were the ones who needed to be cared for, which was the reason they were there in the first place. Perhaps it's because it was a space where people were really listening to each other. The health coaches didn't have an agenda. They weren't marketing a wellness product. They truly wanted to meet the participants where they were at, with kindness.

It's also this attitude of abundance, Dr. Markle said, that contributes to people wanting to volunteer to be peer leaders once they've graduated from the program and to give back by helping future students. One student who graduated told Dr. Markle that being at Open Source Wellness was like encountering a "life force starter engine." "And I love that and I think part of what we're trying to get our coaches to do is to be like a generous starter engine," she said. "It really is this sense of like, if we can be generous, if we can create abundance, people are going to want to contribute to each other and that's what we're banking on."

In another anecdote Dr. Markle shared with me, a participant said she felt the "value" in the program was reciprocated "in all interactions" and exchanges. But it's not just anecdotal evidence that provides testament that what Dr. Markle has created is working. She has reported positive clinical outcomes, such as reductions in depression, anxiety, social isolation, and blood pressure, and significant increases in physical activity and fruit and vegetable consumption. Most notably, she said, there has been a 77 percent reduction in emergency department visits and unplanned hospitalizations among participants—all because most of their doctors prescribed them a community that cared for them.

My visit to Open Source Wellness stuck with me for a long time while I worked on this book. In a society where medical insurance isn't even a human right, it was so refreshing to see accessible wellness in an environment where people actually cared for each other. While acts of kindness and generosity are sure to be catalysts for improving health and societal resilience, as they're shown to do in the wake of a crisis, it occurred to me that this only works when the generosity is welcomed and reciprocated—in other words, when the foundation of the exchange exists in a system that prioritizes abundance.

12 *When You Can't Give, Witness*

During the COVID-19 pandemic, Alan Ross, a professor at Haas Business School at the University of California, Berkeley, took a walk around the Berkeley Hills. At the time, much of life felt glum, he told me in an interview. But one day on his daily walk he saw a young girl who had set up a cookie stand. Initially he appropriately thought "how sweet." This girl is selling cookies. He could have moved on with his day, but luckily he noticed the sign. The cookies weren't for sale. They were free. This pulled at Ross's heartstrings. "Here during these dark days," he recalled, "she was this shining light."

It's not that Ross is a stranger to kindness, but as a business professor he might be more attuned to a freebie than most.

In fact, for years Ross has taught a required class for students at the business school on business ethics. Over the course of his time at Berkeley, he has noticed there has been an uptick in gratitude in business culture and well-being. But as he is quick to note, one can be grateful and "still a jerk." This might be why on the first day of his class he starts by asking his students, what is your responsibility to society? The average student responds by stating that they vote. "The Supreme Court has ruled that corporations basically are people now," he says, arguing that therefore corporate social responsibilities are no different from a citizen's social responsibility. "And it's amazing how little the average student feels they owe to society to give," he said.

When Ross's parents died, he inherited some money. Taking a page from his own teachings, he asked himself, what do I owe to society? Hesitant to spend the money on himself, he decided he wanted to spend it on a project that would give back to his community. He wasn't quite sure what that was until he saw that little girl giving away cookies. "I thought this kid should have a spotlight shined on her," he said. "What she's doing is really special and makes our community so much greater."

Around the same time, a dorm had been built at Berkeley and named Slottman Hall, after the late professor emeritus William Slottman. At Berkeley, Ross explained, having a dorm named after you is the only honor you can receive without making a financial contribution to the school. Slottman's memorial reminded Ross of a man named Chris Walton, a preschool teacher who believed in the power of kindness and who taught Ross's own kids when they were young. All of this led to Ross's initiative today: the Chris Kindness Awards. Each month, a small community in Berkeley nominates people for their big or small acts of kindness. The only requirement is that the nominees live, work, or go to school in Berkeley. At the end of the month, a small panel of judges picks three finalists and the community then votes for their favorite. The winner is awarded $1,000, which they can keep, share, or donate to a charitable cause.

The first winner of the Chris Kindness Awards was a woman named Michele Williams. The person who nominated her was the parent of a child with special needs, who did not have many friends.

But the one friend who reached out every other month to get together with him was Williams. "With kids of her own and all week spent teaching other children in BUSD, she still regularly asks my child for a Peets meet-up to hear how he is doing and catch up," the nominator said.[1]

Another winner of the Chris Kindness Awards was a woman named Carmen Garcia, who worked at a Mobil gas station in Berkeley. The person who nominated her said that when they relocated to Berkeley, they were emerging from a traumatic year of loss. They were starting their life over, alone. One day on their way to work, when they stopped at the Mobil station to grab a drink, Garcia gave them a hug. "Carmen somehow saw right through me, and her empathy made me feel welcome and safe," the nominator said.[2] "Her kindness breaks through the plexiglass as she makes every customer feel like a friend." The nominator, who shared a holiday meal with Garcia and helped make some edits to her daughter's college essay, said she always thought she was Garcia's "favorite" customer. But then she learned there were more. Garcia has made an effort to get to know her customers, and when they're having a bad day, she comes out and just gives them a hug.

Ross said he hasn't been able to get to know each winner and nominee outside of the Chris Kindness Awards. But he's curious to know how they were raised and why they're motivated to be extraordinarily kind in a world that encourages us to be biased toward our own self-interests. These people think differently, Ross insisted, adding that there's nothing superficial or forced about their kind deeds. As a business professor, Ross has brought in some of the winners to talk to his students. He hears his students talking about starting salary as something significant in life, but he's trying to encourage them to see beyond profits and making money. "It really saddens me, the more I get involved with this, to see how we're cheating ourselves by thinking we're spending our time wisely," he said. "And we're not, not as individuals and as a society."

Ross said he has experienced a personal shift just by being in the presence of people who have seemingly prioritized being kind to others on a regular basis. "Being around them makes others so happy and want to do acts of kindness," he said. "I think it's contagious."

In 1686, Sir Isaac Newton first presented his three laws of motion in his book *Philosophiae Naturalis Principia Mathematica*.[3] His first law stated that resting objects remain still unless they're disturbed by an unbalanced force; for instance, like how a dog sleeps until something wakes it up. The second law was a mathematical equation to quantify the net force of an object: $F = ma$ (force equals mass times acceleration). The third law stated that for every action there is an equal and opposite reaction. While Newton was talking about the physical world, this theory can also be thought of in terms of what people refer to as karma.

In Western culture, karma is often thought of as "good" or "bad" luck. If you do something good for others, something good will come to you. But this line of thought is based on the idea of scarcity, that is, that goodness is in short supply. But karma is a Sanskrit word meaning *action*. And the truth is, plenty of bad things happen to good people, and no amount of magical thinking can make sense of this. Poet Maggie Smith wrote about this in her popular poem "Good Bones." "For every bird, there is a stone thrown at a bird," Smith wrote. Perhaps what we can learn from Newton's third law of motion in the context of altruism is that an act of generosity doesn't need to be done under the pretext that something will be received in return.

Instead, if we all live with a mindset of abundance, like at Elizabeth Markle's nonprofit, or one which embraces a stewardship mentality, like mālama, we can trust that a good deed will eventually come back to us. It just might not be in a traditionally transactional way. We can trust that an act of kindness doesn't end when the act itself is over. If we trust each other to be kind and cooperate with one another, we can be certain that we will come across an abundance of kindness in our lifetimes. And this sense of knowing will give us the resilience and strength to weather stormy days.

. . .

In the video *Unsung Hero*, a young Thai man's day begins with water pouring on his head.[4] Instead of getting angry about it, he reaches for a nearby potted plant and places it under the leak that got him wet. Next, a hungry dog begs for his lunch. Instead of getting cranky, he obliges. For three minutes, viewers watch this man continue to perform acts of kindness that appear to be at his expense. The narrator asks in Thai, what does he get in return for doing this every day? Nothing, the narrator answers. He won't get richer. He won't appear on television. But what he does receive are emotions, the narrator says. He witnesses other people's happiness and experiences a deeper understanding of the world. He is able to feel more love and other emotions that money can't buy. Strangely enough, the video is an advertisement for Thai Life Insurance. But the impact it has had on people can be measured by the comments section on YouTube. "I watched it more than 1,000 times," one commenter said. "Watching this as a kid gave me hope and really changed my life."

Researchers at UCLA's Bedari Kindness Institute were curious to know more about the ripple effect of witnessing kindness.[5] In one experiment, they had a group of participants watch the *Unsung Hero* video, while another group watched a video of a man performing entertaining acrobatic stunts. For those who participated in person, researchers gave each person five $1 bills as a payment for their time. They also handed them an envelope they could use to make a donation to UCLA Mattel Children's Hospital. Researchers found that participants who viewed the *Unsung Hero* video gave 25 percent more to the children's hospital than those who watched the video of the athletic stunts and concluded this was one example of a "prosocial contagion."

In other words, witnessing kindness can make kindness contagious. You might be thinking, all of this insight from a life insurance commercial in Thailand? But in Dacher Keltner's book *Awe: The New Science of Everyday Wonder and How It Can Transform Your Life*, he explained the power and new science behind the emotion of awe. "Awe is the feeling of being in the presence of something vast that transcends your understanding of the world," Keltner wrote.[6] It can occur after watching your child take their first steps, viewing the vastness of the mountains, or even seeing Beyoncé in concert. Despite our associating awe with dramatic and once-in-a-lifetime events, awe can also be found by witnessing the kindness of others. That's likely why the Thai commercial had such an impact on so many people.

In one of Keltner's studies, he and his colleagues collected over 2,600 narratives from people, spanning 20 different languages, on their experiences with awe. Over 95 percent of the stories that inspired awe were about people taking action on behalf of others. The stories ranged from watching people perform CPR to revive a victim to bystanders interrupting crimes. "Within the study of morality, it has long been the view that we find our moral compass in the teaching of abstract principles, the study of great texts, or the leadership of charismatic gurus and great sages," Keltner wrote.[7] "In fact, we are just as likely to find our 'moral law within' in the awe we feel for the wonders of others nearby."

Dr. Michael Amster, a doctor and coauthor of the book *The Power of Awe*, told me in an interview that awe is something that broadly "transcends our perception" of the world. He said when people are experiencing awe, they are connected to something greater than the "small self." "Witnessing acts of kindness is an incredible conduit to experiencing awe," he said. "When you witness acts of kindness, you can experience awe, and by practicing those acts of kindness ourselves, we feel that connection with somebody else."

Dr. Amster has worked with doctors and nurses. In his research, he has found that awe can have protective effects against burnout. In America, many caregivers—parents, adult children caring for their older parents, or people caring for a sick family member or friend—are experiencing burnout. According to the National Alliance for Caregiving (NAC) and the 2020 AARP report titled *Caregiving in the U.S.*, an estimated 41.8 million Americans provide unpaid care to adults over the age of 50.[8] These people, according to a Blue Cross Blue Shield report on the impact of caregiving, are at a higher risk for various mental and physical health problems, such as anxiety, depression, obesity, and hypertension. If caring truly has all the health benefits I've written about, why is caregiving in America putting people's health in a precarious situation? As Dr. Steve Cole told me, it's because of the way we approach caregiving. It's isolating, taxing, and often done for profit. It's done in such a way that caregivers never have a moment to refill their cups, due to a lack of social and systemic support.

Burnout, Dr. Amster said, happens when a person is giving and giving. When it's become nothing but a job. When a person has lost the sense of human connection, the precise factor that evokes awe. When caring is task oriented. The antidote, he said, could be to find the awe in it again. To experience the benefits of awe, let's say as a buffer against burnout, Dr. Amster has a three-step approach to experience it in ordinary ways. With the acronym AWE, the first letter stands for *attention*, which means to bring your full, undivided attention to the moment you're in and the person you're connecting with in the moment. "It's a unique opportunity to be in awe of that life experience, of being alive and human in that moment," Dr. Amster told me. The *W* stands for *wait*, which Dr. Amster says is a moment where the person can pause and give themselves some time to digest and soak in the experience. Finally, the *E* stands for *exhale and expand*. "The vagus nerve is attached to the very bottom of your diaphragm," he said as he took a deep breath. "Even if you just make the sound of 'awe' right now, with a nice inhale in and the long exhale out, I just feel immediately more calm and present." When a caregiver can have a moment to pause and reflect, and be like "'Wow, that was like an awesome connection with that patient,' I think there's a mindset shift," that can ease the burnout, he said.

Most of the time, Dr. Amster added, people are on autopilot in life. They're not really even conscious of what they're doing. Our thoughts are going a million miles a minute. But when a person has a moment of awe, they become the experience of their life, he said.

When it comes to altruism, to being kind and witnessing acts of kindness, there is a connectedness of a shared experience that evokes awe and has its own health benefits.

• • •

In a paper published in the journal *Nature Human Behaviour*, researchers randomly analyzed 105,000 headlines and 370 million impressions from a dataset of articles published by the online news website Upworthy.[9] In their analysis, the researchers found that each negative word increased the click-through rate by more than 2 percent. "The presence of positive words in a news headline significantly decreases the likelihood of a headline being clicked on," the researchers said. While negative news is likely to be more popular, that doesn't mean it's good for public health. In 2020, the American Psychological Association reported that more than 50 percent of people surveyed said the news causes them stress.[10] Many reported feelings of anxiety, sleep loss, and fatigue. Yet 1 in 10 adults said they check the news every hour.

There is a saying in journalism, "If it bleeds, it leads," meaning that news centered around tragedy makes the front page. But as Derek Thompson wrote in *The Atlantic*, negativity is not only a journalist's problem. "It's a human problem—or, more to the point, a collective-action problem, in a dual-sided marketplace," he wrote.[11] Many say that the news is full of negative stories because humans are wired with a negativity bias, which has created a vicious cycle. Humans are biased to click on bad news and therefore journalists are more likely to write about it. But what if we reversed the cycle? I'm not suggesting misreporting or ignoring the hard truths about this world, but instead shining a little more light on the good.

In a study published in *PLOS One*, researchers were curious to find out what could act as a buffer against bad news.[12] "Given that horrific acts happen and need to be reported, we examined if news stories featuring others' kindness could undo the aversive effects of news stories featuring others' immorality," the researchers posited. The researchers conducted their experiment in two parts. First, they tested whether media exposure to acts of kindness that occurred in response to something bad, like a terrorist attack, could alleviate the negative effects of learning about the terrorist attack. In the second part, they examined more broadly if the adverse effects of negative stories could be alleviated by featuring news stories about acts of kindness, volunteering, or caring for the homeless. In both parts of the study, the researchers found that news consumers were more likely to experience an elevation in their mood, and still believe in the goodness of others, after reading about both the good and the bad. "Given this, we suggest there is merit in journalists shining a light on others' kindness if people's affective well-being and belief in the goodness of humanity is to remain intact," the researchers concluded.

During the peak of the COVID-19 pandemic, Sean Devlin found himself obsessively checking the news. On a daily basis, he followed COVID-19 case counts. Time and time again he gravitated toward the negativity surrounding everything. Soon, it began to take a toll on his mental health. Fortunately, Devlin was able to take a step back and reflect on what was happening. As he talked with friends and family, he realized he wasn't alone—they felt the same way and were feeling overwhelmed by all the negativity in the news. "I wondered if there could be a more healthy alternative," he said to me in an interview. But he questioned if there was even enough "good news" out there to report. Thanks to his background in email marketing and his curiosity about whether good news even existed, he ended up creating what is known today as Nice News. As an online news platform, the email-first publication delivers good news to people's inboxes. "Positively newsworthy" is its slogan.

"People don't realize, and even we didn't realize when we first started doing this, that there's a lot out there to publish on a daily basis," he said. Devlin currently has a small staff of three and publishes a mix of original reports and curated ones. "It's surprising that there's this much content out there that's centered around positivity," he said. Devlin emphasized that they're not trying to replace news organizations, but rather provide readers with a more positive alternative, to act as a buffer on the internet. For this reason, Nice News doesn't cover every single major news story—only ones that are inclusive and uplifting, and can create an "uplifting mindset." "Even if it is like a tragedy, people are coming together to help and it's inspirational because in tragedy there's also kind of an opportunity to bring people together," he said.

As one might expect, the response from readers has been positive. Right now, Nice News has over 450,000 subscribers. "We've grown relatively quickly, over the last year and a half," he said. And part of the reason for that is because a lot of the content they're sharing is really emotional—it strikes a chord with people. People want to share that with their friends and loved ones, he added.

Devlin said he receives emails on a daily basis from people thanking him for sharing positive news. He was kind enough to share a few testimonials with me.

"Just signed up for your daily emails last week, and I already feel an improvement in my quality of life," one reader wrote to him. "I stopped watching the news about 6 years ago when it was reported that a father in Las Vegas threw all four of his children off a bridge; Nice News was the first time in 6 years that I even wanted to read about what was going on in the world," another said. "I have avoided the news all my life (I'm 66) because it just didn't work for or suit me," another reader wrote. "However, with Nice News, I can practice embracing the wonderfulness of this earth and also feel informed/in the loop, rather than avoiding it."

Devlin said the mission of Nice News is ultimately to, in a way, turn on the switch in all of us to have a more positive bias, to give people a sense of hope and possibility. With that kind of mindset, positive action can follow. His ultimate hope is that Nice News can have a tangible effect in people's lives.

"I think when you have an optimistic mindset, you're able to kind of create solutions," Devlin told me. "And in your everyday life, you're able to connect with others and have this realization that there really is a lot of good out there."

Certainly this connects with the idea that kindness is contagious. In thinking about how to build systemic caring, and a culture of caring, it becomes clear to me that witnessing kindness is key. We need a culture where awe can inspire people to have a more positive outlook in life, which will make them feel safe and therefore be kinder and more generous to others.

• • •

After interviewing Alan Ross, I received a special invite from him. He was hosting his first ever anniversary party for the Chris Kindness Awards. Previous award winners would be in attendance, sharing their stories, he said. I was curious to find out how contagious kindness could be in real life. What would be the immediate effects of being in the presence of altruism without having to engage in an altruistic act myself? I arrived on a sunny October Sunday afternoon at Oak Live Park in Berkeley. I crossed a small wooden bridge over a stream and went through a miniature redwood forest. I found Ross standing behind a white folding table. I found it a bit amusing that a business professor was wearing a shirt that read "Kindness pays." I was also surprised by how many people were in attendance for an event centered around kindness in the community. It was likely the biggest gathering in the park that day. After people ate and enjoyed the music (yes, there was live music, appropriately provided by Berkeley student volunteers), Ross kicked off the program by sharing with everyone how he started the initiative. Next, the first recipient of the award, Michele Williams, spoke and shared how a month before she received the award her father passed away. She revealed she ended up using the award money to pay it forward in honor of her father. Another Chris Kindness Award recipient, Egbert Villegas, was nominated because he saw a car flipped over on the highway and pulled over to help the person who was still in the driver's seat. When asked to speak, he said he didn't really know what to say. "We're all here to celebrate the kindness that surrounds us," he said. "All of you who do acts of kindness, know that it doesn't go unnoticed." These were words he embodied.

Another recipient named Chris Little was nominated by a friend from church who noticed his "kind way of being." In front of the crowd, Little said he was taken aback when he was nominated because he didn't feel like he did anything big. "Believe me, I have a lot of trouble being nice sometimes," he said in all seriousness. But he did have a few thoughts to share. Kindness, Little said, starts at home. "It takes little to no effort to make someone's day, almost every day," he said. "I don't have money or riches, but I do try to be nice and help out where I can." Another man named Anthony, who was a finalist for the award this month, shared how his ability to help another person wouldn't have been possible without a previous act of kindness he received. He and his family were previously unhoused, but his grandmother left them a small inheritance. His mom was able to use this money to turn their whole lives around by getting a house. Next, this month's winner, Kevin Adler, founder of the nonprofit Miracle Messages, took the stage. "I'm just imagining if every park in the world had a Chris Kindness Award going on," he said. "It would be a much better place."

Adler said he first became interested in the issue of homelessness when his uncle lived on the streets for 30 years. Adler said he never looked at him as a homeless man because he was family. It wasn't until his uncle passed away that he realized he had a shift in how he saw homeless people. They were people who needed to be loved, not problems to be solved. In other words, they needed to be seen and treated as human beings. As a result, Adler spent a year getting to know his homeless neighbors and asked some of them to wear GoPro cameras and narrate what their experience was like. What he saw, he said, was heartbreaking. Then one clip he heard changed his life. "I never realized I was homeless when I lost my housing, only when I lost my family and friends," an unhoused person said.

That insight led him to his work on family reunification and the phone buddy program, where 350 unhoused people share weekly phone calls and text messages with volunteers. Today, Adler is working on a basic income housing project. While it's still early in the program, they've found that two-thirds of people who received

$500 every month for six months were able to secure stable housing. They've now expanded the program.

I have no way to measure how kind people were after witnessing so much altruism that day. But what I can say is that I noticed a lot of themes at the event: people seemed more open, socially, than at other events not centered around kindness. I had two strangers approach me and ask how I was connected to the event. People were very willing to share information and resources related to my book. After hearing the stories of award recipients, I do know that for me, the event left me feeling inspired for the rest of the day and week. It also made me want to do something kind, or at least carry on with my day with a more generous mindset.

For instance, when I came home, I walked in to find my husband embarking on a journey of a dinner recipe that would take hours to cook. I didn't want to ruin his kind gesture. I knew he was trying to do something nice. But usually, in this situation, starting an ambitious recipe at 6 p.m. on a Sunday night meant a later cleanup and later bedtime, which I couldn't stand with an infant who still woke up during the night.

Instead of complaining about him making dinner, I made a conscious effort to proceed with kindness. As one Chris Kindness Award recipient said, kindness starts at home. I chose to find humor in the situation. The warm-hearted feelings carried over to the next day. In reflecting on my Sunday, I realized that the event also evoked another emotion in me—awe. Certainly that was part of the reason the event stood out in my memory weeks later.

A couple of months after the Berkeley event, I interviewed Dr. Emiliana Simon-Thomas, science director of the Greater Good Science Center. Anecdotally, I asked if there was any science to support the notion that there are benefits to being a witness to kindness, to being a third party, removed from the exchange. Simon-Thomas pointed me to a study by Joseph Chancellor and Sonja Lyubomirsky, published in the journal *Emotion*.[13] In the study, the researchers assigned one group of participants to perform random acts of kindness, one group to receive acts of kindness, and a third to simply witness kindness. All three groups benefited. Those who simply witnessed had lower levels of loneliness, lower levels of depression, and more positive emotions. And it turned out there was a term to describe what people feel as a witness of caring: *moral elevation*, which is the emotional response to witnessing the virtuous deeds of others.

Similar to being wired to care for others, we are wired to be inspired by witnessing care. "Encountering moral beauty is fundamentally rewarding," Simon-Thomas said. "Our nervous system signals that as pleasure, as something we want to approach and pursue again or emulate ourselves."

With all the suffering in the world, not everyone is in a place to be kind, to give, to share, and to care. We are human, after all. We aren't biblical saints or angels. We won't always easily stretch in the direction of kindness and embrace it. Sometimes our cups are too empty for us to even think about others. Just as we winter, I like to think that throughout our lives we find ourselves in seasons of giving, seasons of receiving, and seasons of witnessing. And among the three, there's a lot of overlap. Sometimes, even if you can only receive, you're giving others the gift of being able to help you. And when you can give, you're giving yourself the gift of giving. And even if all you can do is witness, you're going to feel more inspired and relaxed, and have a stronger sense of feeling safe in society. It's how we build cooperation. It's how we build trust. It's how we build resilience. It's how we crawl out of our modern loneliness crisis. And it's how we build a better future for us all.

"We're an ultrasocial species," Simon-Thomas said. "These moral aptitudes, this just code of fairness, of equity, of justice is really deep in our nervous system."

Epilogue

Four months postpartum, I signed a contract for this book. In other words, I was in a season of extreme giving, that is, in giving everything I had in my heart and soul to one tiny human. In fact, I had never given so much of myself as I gave to my daughter. For anyone else in my orbit, I had very little to give. To them, I probably looked a lot like I was in a season of receiving, as in I was taking all the help, support, and very limited moments to myself that I could get. As one mom in my new moms group said, "Having a baby is like having a bomb go off in your life." For many (not all), having a baby isn't a crisis, but the aftermath is certainly similar to one. It instantly makes you realize what matters most in your life by snapping you back into the moment. It forces you to reevaluate your priorities and cast trivial distractions aside. It's easy to find yourself in survival mode, especially as a woman who traditionally carries the burden of caregiving.

Fortunately, I had a lot of support, which not everyone in this world has in this vulnerable period of life. And at the same time, I faced plenty of anxieties, loneliness, and sometimes even despair. At that moment, if someone told me to go out and volunteer, I would have told them they were crazy. In fact, this wouldn't have helped me at all. It's this tension that I've struggled with since the first day of tackling this subject. How can I tell 53 million caregivers in America, most of whom are women, who are angry at the system, who aren't only mothers but caregivers in every form, that the antidote is what's burning them out in the first place?

If there's one major takeaway from this book, it's that the benefits of altruism work along a dose-response curve and run in tandem with the season of life you're in—giving, receiving, or witnessing. There's no one-size-fits-all solution when it comes to experiencing the benefits of altruism. This could be why not all research on using altruism as a health intervention yields positive results—like one study that found acts of kindness made no difference to the giver's well-being (though it was linked to a reduction in loneliness).[1] I think the outcome and results vary depending on where people are in their lives, and that's where self-kindness comes in. We need to first and foremost be kind to ourselves and the capacities we have to give. Caring for others as self-care won't work for us if we are constantly thinking that what we can give isn't enough, or if we simply aren't in a season of giving.

With that in mind, I think it's vital to emphasize that caring itself isn't the source of burnout in our society. The problem is we are caring in a culture that forces us to do so in isolation, leading to the phenomenon of burnout. As I've reported, caring for others and giving to others is good for our physical, mental, and emotional health in a myriad of ways. It can be energizing. It can help us live longer. It can stop disease from getting worse. It is a way out of loneliness, which is just as detrimental to our health as smoking. Yet sometimes it can feel impossible. When you find yourself in a place of impossibility, know that if other people are prioritizing caring for others, you'll still benefit by being a witness. And when your season of giving comes around again, that will be your chance to give and inspire someone else when they're in a season of witnessing. I also believe that even if you feel like you have nothing to give, that's simply not true. Even in your darkest moments, the tiniest gesture of kindness shown to someone else will briefly alleviate the pain in your struggle.

However, if you are feeling lonely and in a season of giving, I hope this book nudges you to rethink what you need to do to escape that feeling of loneliness. Perhaps the solution requires a shift in mindset to the paradox I've been promoting all along, which is that self-care can be other care. If you're in a season of receiving, know that being vulnerable and asking for help is also a form of giving, because you're giving others the gift of helping you. Asking for help doesn't make you a burden. It's giving others the opportunity to participate in a human exchange that's been key to the survival of our species since the beginning of time.

If you're a retiree, let's say, instead of taking tennis lessons, consider finding a weekly volunteer gig that will not only keep you physically active but also connect you to a greater purpose than yourself—something that will lead to social connectedness and a sense of community with people you feel safe around.

As I illustrated at the beginning of the book, time and time again we see that a crisis, of whatever magnitude, has a way of bringing people together. There are both sociological and biological reasons why this happens. Humans have evolved in groups; there truly is power in numbers when it comes to survival. An act of kindness during a crisis can soothe the overly activated nervous system of both the giver and the receiver. In 1971, biologist Robert

Trivers coined the term *reciprocal altruism*, which is the exchange of altruistic acts between individuals "to produce a net benefit on both sides."[2] As I explored at the end of this book, there's a third part to this, which is the witness.

Another item I wanted to address is the significance of intention. Throughout my work on this book, spanning my own journalistic research on the topic for nearly a decade, people have raised concerns about the idea of promoting volunteering and altruism for the benefit of the giver. Can the benefits still exist if this is the reason you're doing good? When I asked Dr. Emiliana Simon-Thomas about this, she said she didn't have a problem with it. If you're also doing it with the intention of helping someone else, what's wrong with knowing that it's going to benefit you, too? Why do we have to be so humble about doing good?

While writing this book, I jotted down bullet points of reminders that I hope you'll find helpful, no matter what season you're in:

- Be generous with your own vitality; just being yourself is an act of generosity to others.
- When it comes to self-care, think about embracing a sense of purpose. Also think of self-care as a way to build resilience, not just a way to feel better.
- A small act of kindness is just as meaningful as a big one.
- When it comes to kindness interventions, three small acts of kindness two days a week are better than one act of kindness every day.
- Embrace an ethos of care in your mindset, like the Mālama Mindset, by seeing yourself as a steward in this world.

- Any volunteering, especially in old age, is better than none. But if you can volunteer weekly, you're setting yourself up for a healthier and longer life. Bonus if you keep in touch with fellow volunteers regularly.
- If you are going to hang out with a friend and go on a hike, consider finding a volunteer activity to do together. This will be the optimal option for your health.
- To evoke wonder in caring for others, practice AWE (focusing your attention, waiting, and exhaling really slows down your sense of time).

Finally, I hope this book serves as a call to action to care more about this world. To live by an ethos of care and kindness. To see caring for one another as a strength, not a weakness.

If you're a manager at a company, some ways to do this are by looking at your paid sick leave and parental leave, and overall, rethinking the ways your company can actually care for the employees and others. If you're a teacher, maybe it's trying to do relationship mapping with your students. Better yet, if you're part of the management at a school, it's finding ways to pay teachers better and give them more support and resources to teach future generations to care about each other. If you're a parent, maybe it's relying on kindness and talking about it with your kid as a buffer against stressful situations. If you're a doctor, maybe it's prescribing volunteering to patients. If you're a journalist, maybe it's trying to write about some of the good news out there as a buffer for the not-so-great news. If you're a politician, maybe it's embracing housing-first policies or universal basic income.

Whoever you are, whatever you do, you have a big role to play. One small act of caring for someone else can change so many lives.

It can give people a small glimmer of hope that there is some good in humanity. It can be the light that drives the darkness out, if only for just a minute. When we give to others, we give to ourselves. Doing good for others will inevitably improve our health as individuals and as a society. It will help us all build resilience because hope is resilience. As a yoga teacher once told me, "Do good, be good."

Notes

Prologue

Epigraph: Rebecca Solnit, *A Paradise Built in Hell: The Extraordinary Communities That Arise in Disaster* (New York: Penguin, 2010), 7.

1. Troy Griggs, K. K. Rebecca Lai, Haeyoun Park, Jugal K. Patel, and Jeremy White, "Minutes to Escape: How One California Wildfire Damaged So Much So Quickly," *New York Times*, October 12, 2017, https://www.nytimes.com/interactive/2017/10/12/us/california-wildfire-conditions-speed.html.

2. Eric Holthaus, "The Firestorm Ravaging Northern California Cities, Explained," *Mother Jones*, October 10, 2017, https://www.motherjones.com/environment/2017/10/the-firestorm-ravaging-northern-california-cities-explained/.

3. Jill Tucker, "Santa Rosa Schools Reopen after Fires, Ready to Help Students with Stress," *San Francisco Chronicle*, October 27, 2017, https://www.sfchronicle.com/education/article/Wildfire-danger-is-past-but-stress-can-linger-12312685.php.

4. Most destructive at the time; more destructive fires would follow years later.

Chapter One

1. California Department of Fish and Wildlife, "Science: Wildfire Impacts," accessed October 18, 2023, https://wildlife.ca.gov/Science-Institute/Wildfire-Impacts.

2. U.S. Forest Service, "First Returners," accessed October 18, 2023, https://www.fs.usda.gov/Internet/FSE_DOCUMENTS/fseprd575963.pdf.

3. Clarke A. Knight, Lysanna Anderson, M. Jane Bunting, et al., "Land Management Explains Major Trends in Forest Structure and Composition over the Last Millennium in California's Klamath Mountains," *Proceedings of the National Academy of Sciences* 119, no. 12 (March 14, 2022): e2116264119, https://doi.org/10.1073/pnas.2116264119.

4. Katherine May, *Wintering: The Power of Rest and Retreat in Difficult Times* (Waterville, ME: Thorndike Press, 2021).

5. In this book, I frequently use the words *altruism, kindness, caring,* and *generosity* interchangeably. It's a stylistic choice to use different words to describe actions or behavior that benefit someone else's well-being. When necessary, I specifically define each word.

6. Rebecca Solnit, *A Paradise Built in Hell: The Extraordinary Communities That Arise in Disaster* (New York: Penguin, 2010).

Chapter Two

1. Nicole Karlis, "Before the Pandemic, They Were Introverts. Now They Aspire to Live More Extroverted Lives," *Salon.com*, June 6, 2021, https://www.salon.com/2021/06/06/introverts-post-pandemic/.

2. National Institute on Alcohol Abuse and Alcoholism (NIAAA), "Alcohol-Related Deaths, which Increased during the First Year of the COVID-19 Pandemic, Continued to Rise in 2021," April 12, 2023, https://www.niaaa.nih.gov/news-events/research-update/alcohol-related-deaths-which-increased-during-first-year-covid-19-pandemic-continued-rise-2021.

3. Nicole Karlis, "Why 'Social Distancing,' if Done Wrong, Can Make You More Vulnerable," *Salon.com*, March 15, 2020, https://www.salon.com/2020/03/15/why-social-distancing-if-done-wrong-can-make-you-more-vulnerable/.

4. Vladimir G. Simkhovitch, "Mutual Aid a Factor of Evolution, by P. Kropotkin," *Political Science Quarterly* 18, no. 4 (December 1903): 702–5, https://doi.org/10.2307/2140787.

5. Jim Angel, "The 1995 Heat Wave in Chicago, Illinois," State Climatologist Office for Illinois, accessed October 18, 2023, https://www.isws.illinois.edu/statecli/general/1995chicago.htm.

6. Qijun Han and Daniel R. Curtis, "Social Responses to Epidemics Depicted by Cinema," *Emerging Infectious Diseases* 26, no. 2 (2020): 389–94, http://doi.org/10.3201/eid2602.181022.

7. Alan Taylor, "Photos: The Volunteers," *The Atlantic*, April 2, 2020, https://www.theatlantic.com/photo/2020/04/photos-the-volunteers/609149/.

8. E. L. Quarantelli, "Disaster Studies: An Analysis of the Social Historical Factors Affecting the Development of Research in the Area," *International Journal of Mass Emergencies and Disasters* 5 (1987): 285–310, http://udspace.udel.edu/handle/19716/1335.

9. Matěj Bělín, Tomáš Jelínek, and Štěpán Jurajda, "Preexisting Social Ties among Auschwitz Prisoners Support Holocaust Survival," *Proceedings of the National Academy of Sciences* 120, no. 29 (2023): e2221654120, https://doi.org/10.1073/pnas.

10. Maritime Museum of the Atlantic, "Halifax Explosion," accessed October 18, 2023, https://maritimemuseum.novascotia.ca/what-see-do/halifax-explosion.

11. Samuel Henry Prince, "Catastrophe and Social Change: Based upon a Sociological Study of the Halifax Disaster" (PhD diss., Columbia University, New York, 1920), https://www.gutenberg.org/files/37580/37580-h/37580-h.htm.

12. Alejandro Portes, "Social Capital: Its Origins and Applications in Modern Sociology," *Annual Review of Sociology* 24 (1998): 1–24.

13. National Weather Service, "Hurricane Andrew: 30 Years Later," accessed October 18, 2023, https://www.weather.gov/lmk/HurricaneAndrew30Years; National Park Service, "Hurricane Andrew," accessed October 18, 2023, https://www.nps.gov/articles/hurricane-andrew-1992.htm.

14. Florida Climate Center, "Sea Level Rise," accessed October 18, 2023, https://climatecenter.fsu.edu/topics/sea-level-rise.

15. Adam M. Straub, Benjamin J. Gray, Liesel Ashley Ritchie, and Duane A. Gill, "Cultivating Disaster Resilience in Rural Oklahoma: Community Disenfranchisement and Relational Aspects of Social Capital," *Journal of Rural Studies* 73 (January 2020): 105–13.

16. Celina Ortiz and Jason Swinderman, "Eusocial and Colony Behavior in Ants," Reed College, Biology 342, Fall 2012, https://www.reed.edu/biology/courses/BIO342/2012_syllabus/2012_WEBSITES/COJS_animalBehavior/index2.html.

17. Jason Bittel, "Exploding Ant Rips Itself Apart to Protect Its Own," *National Geographic*, April 19, 2018, https://www.nationalgeographic.com/animals/article/animals-ants-borneo-exploding-defense.

18. Charles Darwin, *The Descent of Man and Selection in Relation to Sex* (New York: D. Appleton, 1871).

19. *Encyclopaedia Britannica*, s.v. "kin selection," accessed November 11, 2023, https://www.britannica.com/topic/kin-selection.

20. *Encyclopaedia Britannica*, s.v. "inclusive fitness," accessed November 11, 2023, https://www.britannica.com/science/inclusive-fitness.

21. George Williams, *Adaptation and Natural Selection* (Princeton, NJ: Princeton University Press, 1966).

22. Felix Warneken and Michael Tomasello, "Altruistic Helping in Human Infants and Young Chimpanzees," *Science* 311 (2006): 1301, https://doi.org/10.1126/science.1121448.

23. Oak Ridge Associated Universities, "Psychosocial Reactions: Phases of Disaster," accessed March 5, 2024, https://www.orau.gov/rsb/pfaird/03-psychosocial-reactions-01-phases-of-disaster.html.

24. U.S. Census Bureau, "New Data Reveal Most Populous Cities Experienced Some of the Largest Decreases," May 26, 2022, https://www.census.gov/library/stories/2022/05/population-shifts-in-cities-and-towns-one-year-into-pandemic.html.

Chapter Three

1. Robert D. Putnam, "Bowling Alone: America's Declining Social Capital," *Journal of Democracy* 6, no. 1 (January 1995): 65–78, https://www.journalofdemocracy.org/articles/bowling-alone-americas-declining-social-capital/.

2. Philip M. Alberti, Carla S. Alvarado, and Heather H. Pierce, "Civic Engagement: A Vital Sign of Health and Democracy," AAMC Center for Health Justice, September 26, 2022, https://www.aamchealthjustice.org/news/polling/civic-engagement.

3. *How to PTA in Challenging Times: Texas PTA Rally Day 2021*, National PTA, 2021, https://www.pta.org/docs/default-source/files/runyourpta/2021/how-we-pta/case-study---texas-pta-rally-day.pdf.

4. Editorial Board, "The Worst Voter Turnout in 72 Years," *New York Times*, November 11, 2014, https://www.nytimes.com/2014/11/12/opinion/the-worst-voter-turnout-in-72-years.html.

5. E. J. Dionne, Norman J. Ornstein, and Thomas E. Mann, *One Nation after Trump* (New York: St. Martin's Press, 2017).

6. U.S. Surgeon General, *Our Epidemic of Loneliness and Isolation: The U.S. Surgeon General's Advisory on the Healing Effects of Social Connection and Community*, 2023, https://www.hhs.gov/sites/default/files/surgeon-general-social-connection-advisory.pdf.

7. Steve Jobs, "Steve Jobs Introducing the iPhone at MacWorld 2007," YouTube, January 9, 2007, video, 14:00, https://www.youtube.com/watch?v=x7qPAY9JqE4.

8. Andrew Perrin and Sara Atske, "About Three-in-Ten U.S. Adults Say They Are 'Almost Constantly' Online," Pew Research Center, March 26, 2021, https://www.pewresearch.org.

9. Nick Statt, "The Creators of the iPhone Are Worried We're Too Addicted to Technology," *The Verge*, June 29, 2017, https://www.theverge.com/2017/6/29/15893960/apple-iphone-creators-smartphone-addiction-ideo-interview.

10. Josh Howarth, "Alarming Average Screen Time Statistics (2023)," *Exploding Topics* (blog), January 13, 2023, https://explodingtopics.com/blog/screen-time-stats.

11. R. Sturm and D. A. Cohen, "Free Time and Physical Activity among Americans 15 Years or Older: Cross-Sectional Analysis of the American Time Use Survey," *Preventing Chronic Disease* 16 (September 26, 2019), http://dx.doi.org/10.5888/pcd16.190017.

12. Screen Time Action Network, Letter to Jessica Henderson Daniel, PhD, ABPP, President, American Psychological Association, August 8, 2018, https://screentimenetwork.org/apa?eType=EmailBlastContent&eId=5026ccf8-74e2-4f10-bc0e-d83dc030c894.

13. Haley Sweetland Edwards, "You're Addicted to Your Smartphone. This Company Thinks It Can Change That," *Time*, April 12, 2018, https://time.com/5237434/youre-addicted-to-your-smartphone-this-company-thinks-it-can-change-that/.

14. Melinda Wenner Moyer, "Kids as Young as 8 Are Using Social Media More than Ever, Study Finds," *New York Times*, March 24, 2022, https://www.nytimes.com/2022/03/24/well/family/child-social-media-use.html.

15. John T. Cacioppo and William Patrick, *Loneliness: Human Nature and the Need for Social Connection* (New York: W. W. Norton, 2008).

16. Fay Bound Alberti, *A Biography of Loneliness: The History of an Emotion* (Oxford: Oxford University Press, 2019).

17. Henry David Thoreau, *The Journal of Henry David Thoreau, 1837–1861*, ed. Damion Searls (New York: New York Review Books Classics, 2009).

18. Dan Russell, Letitia Anne Peplau, and Mary Lund Ferguson, "Developing a Measure of Loneliness," *Journal of Personality Assessment* 42, no. 3 (July 1978): 290–94, http://dx.doi.org/10.1207/s15327752jpa4203_11.

19. "New Cigna Study Reveals Loneliness at Epidemic Levels in America: Research Puts Spotlight on the Impact of Loneliness in the U.S. and Potential Root Causes," Cigna, May 1, 2018, https://www.multivu.com/players/English/8294451-cigna-us-loneliness-survey/.

20. Harvard Graduate School of Education and Making Caring Common Project, "Loneliness in America: How the Pandemic Has Deepened an Epidemic of Loneliness and What We Can Do about It," February 2021, https://mcc.gse.harvard.edu/reports/loneliness-in-america.

21. Nana Baah, "Young People Are Lonelier than Ever," *Vice*, April 22, 2022, https://www.vice.com/en/article/z3n5aj/loneliness-epidemic-young-people.

22. Centers for Disease Control and Prevention, "Loneliness and Social Isolation Linked to Serious Health Conditions," last reviewed April 29, 2021, https://www.cdc.gov/aging/publications/features/lonely-older-adults.html.

23. American Heart Association News, "Social Isolation, Loneliness Can Damage Heart and Brain Health, Report Says," August 4, 2022, https://www.heart.org/en/news/2022/08/04/social-isolation-loneliness-can-damage-heart-and-brain-health-report-says.

Chapter Four

1. Cleveland Clinic, "Cranial Nerves," last reviewed October 27, 2021, https://my.clevelandclinic.org/health/body/21998-cranial-nerves.

2. While I refer to the vagus nerve in the singular, it's technically a pair of nerves. *Physiopedia*, s.v. "vagus nerve," https://www.physio-pedia.com/Vagus_Nerve.

3. Alok Jha, "Why Crying Babies Are So Hard to Ignore," *The Guardian*, October 17, 2012, https://www.theguardian.com/science/2012/oct/17/crying-babies-hard-ignore.

4. Jason G. Goldman, "Ed Tronick and the 'Still Face Experiment,'" *Scientific American*, October 18, 2010, https://blogs.scientificamerican.com/thoughtful-animal/ed-tronick-and-the-8220-still-face-experiment-8221/.

5. Polyvagal Institute, "What Is the Polyvagal Theory?," https://www.polyvagalinstitute.org/whatispolyvagaltheory.

6. Refinery29, "I Live in Los Angeles, Make $75,000 a Year, and Spent $2,003 on My Wellness Routine This Week," *Feel Good Diaries*, June 17, 2019, https://www.refinery29.com/en-us/2019/06/235357/wellness-routine-barrys-bootcamp-smoothies.

7. *Field Agent*, "Millennials, Boomers, and 2015 Resolutions: 5 Key Generational Differences," January 13, 2015, https://blog.fieldagent.net/millennials-boomers-new-years-resolutions-5-key-generational-differences.

8. Nicole Karlis, "How the 2010s Became the Decade of Self-Care," *Salon.com*, December 21, 2019, https://www.salon.com/2019/12/21/how-the-2010s-became-the-decade-of-self-care/.

9. Philip Brickman and Donald T. Campbell, "Hedonic Relativism and Planning the Good Society," in *Adaptation-Level Theory*, ed. M. H. Appley (New York: Academic Press, 1971), 287–305.

10. Philip Brickman, Dan Coates, and Ronnie Janoff-Bulman, "Lottery Winners and Accident Victims: Is Happiness Relative?," *Journal of Personality and Social Psychology* 36, no. 8 (1978): 917–27, https://gwern.net/doc/psychology/1978-brickman.pdf.

11. McKinsey & Company, "Feeling Good: The Future of the $1.5 Trillion Wellness Market," April 8, 2021, https://www.mckinsey.com/industries/consumer-packaged-goods/our-insights/feeling-good-the-future-of-the-1-5-trillion-wellness-market.

12. S. Alexander Haslam, Charlotte McMahon, Tegan Cruwys, Catherine Haslam, Jolanda Jetten, and Niklas K. Steffens, "Social Cure, What Social Cure? The Propensity to Underestimate the Importance of Social Factors for Health," *Social Science and Medicine* 198 (2018): 14–21, https://doi.org/10.1016/j.socscimed.2017.12.020.

13. S. W. Porges and S. A. Furman, "The Early Development of the Autonomic Nervous System Provides a Neural Platform for Social Behavior: A Polyvagal Perspective," *Infant and Child Development* 20, no. 1 (January/February 2011): 106–18, https://doi.org/10.1002/icd.688.

14. Galway Kinnell, "Saint Francis and the Sow," in *Mortal Acts, Mortal Words* (Boston: Houghton Mifflin, 1980), available at https://www.encyclopedia.com/arts/educational-magazines/saint-francis-and-sow.

Chapter Five

1. Lisa Luckett, "Love vs Fear. Can We See 9/11 in a New Light?," TEDxNewBedford, December 7, 2016, https://www.youtube.com/watch?v=LOnMZXbII7M.

2. Nicole Karlis, "Lisa Luckett, 9/11 Widow, Explains How Tragedy Helps Us Grow," *Salon.com*, September 11, 2018, https://www.salon.com/2018/09/11/lisa-luckett-911-widow-explains-how-tragedy-helps-us-grow/.

3. Barbara L. Fredrickson, "What Good Are Positive Emotions?," *Review of General Psychology* 2, no. 3 (1998): 300–319, https://doi.org/10.1037/1089-2680.2.3.300.

4. Sonja Lyubomirsky, "On Studying Positive Emotions," *Prevention and Treatment* 3, no. 1, (March 7, 2000), https://psycnet.apa.org/doi/10.1037/1522-3736.3.1.35c.

5. Zak Stambor, "A Key to Happiness," *Monitor on Psychology* 37, no. 9 (October 2006), https://www.apa.org/monitor/oct06/key.

6. L. E. Alden and J. L. Trew, "If It Makes You Happy: Engaging in Kind Acts Increases Positive Affect in Socially Anxious Individuals," *Emotion* 13, no. 1 (2013): 64–75, https://doi.org/10.1037/a0027761.

7. David R. Cregg and Jennifer S. Cheavens, "Healing through Helping: An Experimental Investigation of Kindness, Social Activities, and Reappraisal as Well-Being Interventions," *Journal of Positive Psychology* 18, no. 6 (2023): 924–41, https://doi.org/10.1080/17439760.2022.2154695.

8. R. M. Ryan and E. L. Deci, "Self-Determination Theory and the Facilitation of Intrinsic Motivation, Social Development, and Well-Being," *American Psychologist* 55 (2000): 68–78.

9. J. Walker, A. Kumar, and T. Gilovich, "Cultivating Gratitude and Giving through Experiential Consumption," *Emotion* 16, no. 8 (2016): 1126–36, https://doi.org/10.1037/emo0000242.

10. Adam Phillips and Barbara Taylor, *On Kindness* (New York: Picador, Macmillan, 2020).

11. WKYC Channel 3, "Boys Do Read: 8-Year-Old Boy Starts Reading Movement in Cleveland," November 23, 2018, https://www.youtube.com/watch?v=aJbTRY2xwfY.

12. Color a Smile, "Thank You's," https://colorasmile.org/thank_you/.

13. NHS England, "Social Prescribing: What Is Social Prescribing?," accessed August 21, 2024, https://www.england.nhs.uk/personalisedcare/social-prescribing/.

14. NHS England, "Social Prescribing—The Power of Time and Connections," https://www.england.nhs.uk/personalisedcare/comprehensive-model/case-studies/social-prescribing-the-power-of-time-and-connections/.

15. 9/11 Day, "Jay Winuk," https://911day.org/leaders/jay-winuk.

Chapter Six

1. University of Pennsylvania, "How Do Natural Disasters Shape the Behavior and Social Networks of Rhesus Macaques?," press release, April 8, 2021, https://penntoday.upenn.edu/news/Penn-neuroscience-natural-disasters-behavior-social-networks-rhesus-macaques.

2. National Weather Service, "Major Hurricane Maria," September 20, 2017, https://www.weather.gov/sju/maria2017.

3. Camille Testard, Sam M. Larson, Marina M. Watowich, Noah Snyder-Mackler, Michael L. Platt, and Lauren J. N. Brent, "Rhesus Macaques Build New Social Connections after a Natural Disaster," *Current Biology* 31, no. 11 (2021): 2299–2309. https://doi.org/10.1016/j.cub.2021.03.029.

4. R. Jones, "It's Good to Give," *Nature Reviews Neuroscience* 7 (2006): 907, https://doi.org/10.1038/nrn2047.

5. Nicole Karlis, "Why Doing Good Is Good for the Do-Gooder," *New York Times*, October 26, 2017, https://www.nytimes.com/2017/10/26/well/mind/why-doing-good-is-good-for-the-do-gooder.html.

6. University of Wisconsin–Madison, "Brain Can Be Trained in Compassion, Study Shows," press release, May 22, 2013, https://news.wisc.edu/brain-can-be-trained-in-compassion-study-shows/.

7. University of Wisconsin–Madison, "Brain Can Be Trained in Compassion."

8. M. T. Johnson, J. M. Fratantoni, K. Tate, and A. S. Moran, "Parenting with a Kind Mind: Exploring Kindness as a Potentiator for Enhanced Brain Health,"

Frontiers in Psychology 13 (2022): 805748, https://doi.org/10.3389/fpsyg.2022.805748.

9. Center for BrainHealth, "The Power of Kindness in Improving Brain Health," press release, *NeuroscienceNews.com*, April 11, 2022, https://neurosciencenews.com/kindness-brain-health-20360/.

10. Hawaii News Now Staff, "Hawaii Saw More than 10M Visitors Last Year, but Not Everyone Is Celebrating," January 30, 2020, https://www.hawaiinewsnow.com/2020/01/31/hawaii-saw-more-than-m-visitors-last-year-not-everyone-is-celebrating/.

11. Data USA, "Hawaii: About," accessed September 1, 2023, https://datausa.io/profile/geo/hawaii.

12. Audrey McAvoy, "Agency to Hawaii Residents: Don't Hate on Tourists," Associated Press, April 24, 2016, https://www.seattletimes.com/nation-world/agency-to-hawaii-residents-dont-hate-on-tourists/.

13. Bishop Museum, "Hidden in Plain Sight: The History of the Healer Stones of Kapaemahu in a Changing Waikīkī," YouTube, streamed live July 8, 2022, presentation, 1:41:00, https://www.youtube.com/watch?v=BK-eNa4kyNg.

14. U.S. Census Bureau, *Remembering Pearl Harbor: A Pearl Harbor Fact Sheet*, https://www.census.gov/history/pdf/pearl-harbor-fact-sheet-1.pdf.

15. Battleship Missouri Memorial, "About Us," accessed September 1, 2023, https://ussmissouri.org/about-us.

Chapter Seven

1. Hiʻipaka LLC, "History of Waimea Valley," https://www.waimeavalley.net/about-waimea.

2. Maui Nui Botanical Gardens, "Koa (Acacia koa)," accessed August 19, 2023, https://mnbg.org/hawaiian-native-plant-collection/koa-acacia-koa.

3. Lorraine Boissoneault, "When Invasive Species Become Local Cuisine," *The Atlantic*, May 19, 2016, https://www.theatlantic.com/science/archive/2016/05/hawaii-invasive-species/483509/.

4. Hawaiʻi Department of Land and Natural Resources, "Rare Plants," https://dlnr.hawaii.gov/ecosystems/rare-plants/.

5. Allan Luks, "Helper's High: The Healing Power of Helping Others," https://allanluks.com/helpers_high.

6. Allan Luks, "Doing Good: Helper's High," *Psychology Today* 22, no. 10 (1988), photocopy of article available at https://ellisarchive.org/sites/default/files/2019-10/Document_20191001_0008.pdf.

7. Lisa Howard, "Volunteering in Late Life May Protect the Brain against Cognitive Decline and Dementia," UC Davis Health News, July 20, 2023, https://health.ucdavis.edu/news/headlines/volunteering-in-late-life-may-protect-the-brain-against-cognitive-decline-and-dementia/2023/07.

8. Eric S. Kim and Sara H. Konrath, "Volunteering Is Prospectively Associated with Health Care Use among Older Adults," *Social Science and Medicine* (January 2016): 122–29, https://doi.org/10.1016/j.socscimed.2015.11.043.

Chapter Eight

1. Amy Yotopoulos, "Three Reasons Why People Don't Volunteer, and What Can Be Done about It," Stanford Center on Longevity, https://longevity.stanford.edu/three-reasons-why-people-dont-volunteer-and-what-can-be-done-about-it/.

2. *Online Etymology Dictionary*, s.v. "volunteer," https://www.etymonline.com/word/volunteer.

3. City of Novato, "City Partners with Dominican University to Develop Reimagining Citizenship Program," January 10, 2018, https://www.novato.org/Home/Components/News/News/5610/637.

4. History.com, "Civilian Conservation Corps," updated March 31, 2021, https://www.history.com/topics/great-depression/civilian-conservation-corps#section_6.

5. California Volunteers, "Statewide Pledge Ceremony," YouTube, October 17, 2023, video, 2:19, https://www.youtube.com/watch?v=hop2IsYrMdE.

Chapter Nine

1. Viktor E. Frankl, *Man's Search for Meaning* (Boston: Beacon Press, 2006).

2. Michele Dillon, *In the Course of a Lifetime* (Berkeley: University of California Press, 2007).

3. Centers for Disease Control and Prevention, "*Pneumocystis* Pneumonia—Los Angeles," *MMWR: Morbidity and Mortality Weekly Report* 30, no. 21

(June 5, 1981): 1–3, https://www.cdc.gov/mmwr/preview/mmwrhtml/june_5.htm.

4. Steve W. Cole, Louise C. Hawkley, Jesusa M. Arevalo, Caroline Y. Sung, Robert M. Rose, and John T. Cacioppo, "Social Regulation of Gene Expression in Human Leukocytes," *Genome Biology* 8, no. 9 (2007): R189, https://doi.org/10.1186/gb-2007-8-9-r189.

5. Steven W. Cole, John T. Cacioppo, Stephanie Cacioppo, et al., "The Type I Interferon Antiviral Gene Program Is Impaired by Lockdown and Preserved by Caregiving," *Proceedings of the National Academy of Sciences* 118, no. 29 (July 16, 2021): e2105803118, https://doi.org/10.1073/pnas.2105803118.

Chapter Ten

1. Attendance Works, "Rising Tide of Chronic Absence Challenges Schools," October 12, 2023, https://www.attendanceworks.org/rising-tide-of-chronic-absence-challenges-schools/?preview=true.

2. Making Caring Common, "Relationship Mapping Strategy," Harvard University Graduate School of Education, accessed November 11, 2023, https://mcc.gse.harvard.edu/resources-for-educators/relationship-mapping-strategy.

3. Making Caring Common, "Virtual Listening Deeply: Strategy and Lesson Plans," Harvard University Graduate School of Education, accessed November 11, 2023. https://static1.squarespace.com/static/5b7c56e255b02c683659fe43/t/5f2afe834450e97f1c74fcb2/1596653206496/Virtual+Listening+Deeply.pdf.

4. Chris Murphy, "Congressional Remarks from Senator Chris Murphy (D-CT)," YouTube, June 14, 2023, speech, 12:47, https://www.youtube.com/watch?v=SAPFkQfjNgI.

5. Levi Boxell, Matthew Gentzkow, and Jesse M. Shapiro, "Cross-Country Trends in Affective Polarization," National Bureau of Economic Research, Working Paper Series no. 26669 (January 2020), https://doi.org//10.3386/w26669.

6. Sara H. Konrath, Edward H. O'Brien, and Courtney Hsing, "Changes in Dispositional Empathy in American College Students Over Time: A Meta-Analysis," *Personality and Social Psychology Review* 15, no. 2 (May 2011): 180–98, https://doi.org/10.1177/1088868310377395.

Chapter Eleven

1. B. Marlin, M. Mitre, J. D'amour, et al., "Oxytocin Enables Maternal Behaviour by Balancing Cortical Inhibition," *Nature* 520 (2015): 499–504, https://doi.org/10.1038/nature14402.

Chapter Twelve

1. Annie Hahn, "Michele Williams, Nov. Winner," Chris Kindness Award, November 30, 2022, https://chriskindnessaward.org/michele-williams-nov-winner/.
2. "Carmen Garcia—Feb. Winner," Chris Kindness Award, March 26, 2023, https://chriskindnessaward.org/carmen-garcia-our-march-winner/.
3. NASA Glenn Research Center, "Newton's Laws of Motion," accessed November 11, 2023, https://www1.grc.nasa.gov/beginners-guide-to-aeronautics/newtons-laws-of-motion.
4. Thai Life Channel, "Unsung Hero," YouTube, April 3, 2014, commercial, 3:05, https://www.youtube.com/watch?v=uaWA2GbcnJU.
5. Jessica Wolf, "Is Kindness Contagious?," *UCLA Magazine*, January 5, 2023, https://newsroom.ucla.edu/magazine/bedari-kindness-institute-contagious.
6. Hope Reese, "How a Bit of Awe Can Improve Your Health," *New York Times*, January 3, 2023, https://www.nytimes.com/2023/01/03/well/live/awe-wonder-dacher-keltner.html.
7. Dacher Keltner, "What's the Most Common Source of Awe?," *Greater Good Magazine*, January 24, 2023, https://greatergood.berkeley.edu/article/item/whats_the_most_common_source_of_awe.
8. Rebecca Schier-Akamelu, "2023 Caregiver Burnout and Stress Statistics," A Place for Mom, last updated June 13, 2023, https://www.aplaceformom.com/senior-living-data/articles/caregiver-burnout-statistics.
9. C. E. Robertson, N. Pröllochs, K. Schwarzenegger, et al. "Negativity Drives Online News Consumption," *Nature Human Behaviour* 7 (2023): 812–22, https://doi.org/10.1038/s41562-023-01538-4.
10. American Psychological Association (APA), "Stress in America™ 2020: A National Mental Health Crisis," October 2020, https://www.apa.org/news/press/releases/stress/2020/report-october.

11. Derek Thompson, "Click Here if You Want to Be Sad," *The Atlantic*, March 24, 2023, https://www.theatlantic.com/newsletters/archive/2023/03/negativity-bias-online-news-consumption/673499/.

12. Kathryn Buchanan and Gillian M. Sandstrom, "Buffering the Effects of Bad News: Exposure to Others' Kindness Alleviates the Aversive Effects of Viewing Others' Acts of Immorality," *PLOS One* (May 17, 2023), https://doi.org/10.1371/journal.pone.0284438.

13. J. Chancellor, S. Margolis, K. Jacobs Bao, and S. Lyubomirsky, "Everyday Prosociality in the Workplace: The Reinforcing Benefits of Giving, Getting, and Glimpsing," *Emotion* 18, no. 4 (2018): 507-17, https://doi.org/10.1037/emo0000321.

Epilogue

1. Megan M. Fritz, Lisa C. Walsh, Steven W. Cole, Elissa Epel, and Sonja Lyubomirsky, "Kindness and Cellular Aging: A Pre-registered Experiment Testing the Effects of Prosocial Behavior on Telomere Length and Well-Being," *Brain, Behavior, and Immunity: Health* 11 (February 2021): 100187, https://doi.org/10.1016/j.bbih.2020.100187.

2. Robert L. Trivers, "The Evolution of Reciprocal Altruism," *Quarterly Review of Biology* 46, no. 1 (March 1971): 35-57, http://www.jstor.org/stable/2822435?origin=JSTOR-pdf.

Bibliography

Alberti, Fay Bound. *A Biography of Loneliness: The History of an Emotion.* Oxford: Oxford University Press, 2019.

Alberti, Philip M., Carla S. Alvarado, and Heather H. Pierce. "Civic Engagement: A Vital Sign of Health and Democracy." AAMC Center for Health Justice, September 26, 2022. https://www.aamchealthjustice.org/news/polling/civic-engagement.

Alden, L. E., and J. L. Trew. "If It Makes You Happy: Engaging in Kind Acts Increases Positive Affect in Socially Anxious Individuals." *Emotion* 13, no. 1 (2013): 64–75. https://doi.org/10.1037/a0027761.

American Heart Association News. "Social Isolation, Loneliness Can Damage Heart and Brain Health, Report Says." August 4, 2022. https://www.heart.org/en/news/2022/08/04/social-isolation-loneliness-can-damage-heart-and-brain-health-report-says.

American Psychological Association (APA). "Stress in America™ 2020: A National Mental Health Crisis." October 2020. https://www.apa.org/news/press/releases/stress/2020/report-october.

Angel, Jim. "The 1995 Heat Wave in Chicago, Illinois." State Climatologist Office for Illinois. Accessed October 18, 2023. https://www.isws.illinois.edu/statecli/general/1995chicago.htm.

Attendance Works. "Rising Tide of Chronic Absence Challenges Schools." October 12, 2023. https://www.attendanceworks.org/rising-tide-of-chronic-absence-challenges-schools/?preview=true.

Baah, Nana. "Young People Are Lonelier than Ever." *Vice*, April 22, 2022. https://www.vice.com/en/article/z3n5aj/loneliness-epidemic-young-people.

Battleship Missouri Memorial. "About Us." Accessed September 1, 2023. https://ussmissouri.org/about-us.

Bělín, Matěj, Tomáš Jelínek, and Štěpán Jurajda. "Preexisting Social Ties among Auschwitz Prisoners Support Holocaust Survival." *Proceedings of the National Academy of Sciences* 120, no. 29 (2023): e2221654120. https://doi.org/10.1073/pnas.

Bishop Museum. "Hidden in Plain Sight: The History of the Healer Stones of Kapaemahu in a Changing Waikīkī." YouTube, streamed live July 8, 2022, presentation, 1:41:00. https://www.youtube.com/watch?v=BK-eNa4kyNg.

Bittel, Jason. "Exploding Ant Rips Itself Apart to Protect Its Own." *National Geographic*, April 19, 2018. https://www.nationalgeographic.com/animals/article/animals-ants-borneo-exploding-defense.

Boissoneault, Lorraine. "When Invasive Species Become Local Cuisine." *The Atlantic*, May 19, 2016. https://www.theatlantic.com/science/archive/2016/05/hawaii-invasive-species/483509/.

Boxell, Levi, Matthew Gentzkow, and Jesse M. Shapiro. "Cross-Country Trends in Affective Polarization." National Bureau of Economic Research, Working Paper Series no. 26669, January 2020. https://doi.org//10.3386/w26669.

Brickman, Philip, and Donald T. Campbell. "Hedonic Relativism and Planning the Good Society." In *Adaptation-Level Theory*, edited by M. H. Appley, 287–305. New York: Academic Press, 1971.

Brickman, Philip, Dan Coates, and Ronnie Janoff-Bulman. "Lottery Winners and Accident Victims: Is Happiness Relative?" *Journal of Personality and Social Psychology* 36, no. 8 (1978): 917–27. https://gwern.net/doc/psychology/1978-brickman.pdf.

Buchanan, Kathryn, and Gillian M. Sandstrom. "Buffering the Effects of Bad News: Exposure to Others' Kindness Alleviates the Aversive Effects of Viewing Others' Acts of Immorality." *PLOS One* (May 17, 2023). https://doi.org/10.1371/journal.pone.0284438.

Cacioppo, John T., and William Patrick. *Loneliness: Human Nature and the Need for Social Connection*. New York: W. W. Norton, 2008.

California Department of Fish and Wildlife. "Science: Wildfire Impacts." Accessed October 18, 2023. https://wildlife.ca.gov/Science-Institute/Wildfire-Impacts.

California Volunteers. "Statewide Pledge Ceremony." YouTube, October 17, 2023, video, 2:19. https://www.youtube.com/watch?v=hop2IsYrMdE.

"Carmen Garcia—Feb. Winner." Chris Kindness Award, March 26, 2023. https://chriskindnessaward.org/carmen-garcia-our-march-winner/.

Center for BrainHealth. "The Power of Kindness in Improving Brain Health." Press release, *NeuroscienceNews.com*, April 11, 2022. https://neurosciencenews.com/kindness-brain-health-20360/.

Centers for Disease Control and Prevention. "Loneliness and Social Isolation Linked to Serious Health Conditions." Last reviewed April 29, 2021. https://www.cdc.gov/aging/publications/features/lonely-older-adults.html.

———. "*Pneumocystis* Pneumonia—Los Angeles." *MMWR: Morbidity and Mortality Weekly Report* 30, no. 21 (June 5, 1981): 1–3. https://www.cdc.gov/mmwr/preview/mmwrhtml/june_5.htm.

Chancellor, J., S. Margolis, K. Jacobs Bao, and S. Lyubomirsky. "Everyday Prosociality in the Workplace: The Reinforcing Benefits of Giving, Getting, and Glimpsing." *Emotion* 18, no. 4 (2018): 507–17. https://doi.org/10.1037/emo0000321.

City of Novato. "City Partners with Dominican University to Develop Reimagining Citizenship Program." January 10, 2018. https://www.novato.org/Home/Components/News/News/5610/637.

Cleveland Clinic. "Cranial Nerves." Last reviewed October 27, 2021. https://my.clevelandclinic.org/health/body/21998-cranial-nerves.

Cole, Steven W., John T. Cacioppo, Stephanie Cacioppo, et al. "The Type I Interferon Antiviral Gene Program Is Impaired by Lockdown and Preserved by Caregiving." *Proceedings of the National Academy of Sciences* 118, no. 29 (July 16, 2021): e2105803118. https://doi.org/10.1073/pnas.2105803118.

Cole, Steve W., Louise C. Hawkley, Jesusa M. Arevalo, Caroline Y. Sung, Robert M. Rose, and John T. Cacioppo. "Social Regulation of Gene Expression in Human Leukocytes." *Genome Biology* 8, no. 9 (2007): R189. https://doi.org/10.1186/gb-2007-8-9-r189.

Color a Smile. "Thank You's." https://colorasmile.org/thank_you/.

Cregg, David R., and Jennifer S. Cheavens. "Healing through Helping: An Experimental Investigation of Kindness, Social Activities, and Reappraisal as Well-Being Interventions." *Journal of Positive Psychology* 18, no. 6 (2023): 924–41. https://doi.org/10.1080/17439760.2022.2154695.

Darwin, Charles. *The Descent of Man and Selection in Relation to Sex*. New York: D. Appleton, 1871.

Data USA. "Hawaii: About." Accessed September 1, 2023. https://datausa.io/profile/geo/hawaii.

Dillon, Michele. *In the Course of a Lifetime*. Berkeley: University of California Press, 2007.

Dionne, E. J., Norman J. Ornstein, and Thomas E. Mann. *One Nation after Trump*. New York: St. Martin's Press, 2017.

Editorial Board. "The Worst Voter Turnout in 72 Years." *New York Times*, November 11, 2014. https://www.nytimes.com/2014/11/12/opinion/the-worst-voter-turnout-in-72-years.html.

Encyclopaedia Britannica. s.v. "inclusive fitness." Accessed November 11, 2023. https://www.britannica.com/science/inclusive-fitness.

———. s.v. "kin selection." Accessed November 11, 2023. https://www.britannica.com/topic/kin-selection.

Field Agent. "Millennials, Boomers, and 2015 Resolutions: 5 Key Generational Differences." January 13, 2015. https://blog.fieldagent.net/millennials-boomers-new-years-resolutions-5-key-generational-differences.

Florida Climate Center. "Sea Level Rise." Accessed October 18, 2023. https://climatecenter.fsu.edu/topics/sea-level-rise.

Frankl, Viktor E. *Man's Search for Meaning*. Boston: Beacon Press, 2006.

Fredrickson, Barbara L. "What Good Are Positive Emotions?" *Review of General Psychology* 2, no. 3 (1998): 300–319. https://doi.org/10.1037/1089-2680.2.3.300.

Fritz, Megan M., Lisa C. Walsh, Steven W. Cole, Elissa Epel, and Sonja Lyubomirsky. "Kindness and Cellular Aging: A Pre-registered Experiment Testing the Effects of Prosocial Behavior on Telomere Length and Well-Being." *Brain, Behavior, and Immunity: Health* 11 (February 2021): 100187. https://doi.org/10.1016/j.bbih.2020.100187.

Goldman, Jason G. "Ed Tronick and the 'Still Face Experiment.'" *Scientific American*, October 18, 2010. https://blogs.scientificamerican.com/thoughtful-animal/ed-tronick-and-the-8220-still-face-experiment-8221/.

Griggs, Troy, K. K. Rebecca Lai, Haeyoun Park, Jugal K. Patel, and Jeremy White. "Minutes to Escape: How One California Wildfire Damaged So Much So Quickly." *New York Times*, October 12, 2017. https://www

.nytimes.com/interactive/2017/10/12/us/california-wildfire-conditions-speed.html.

Hahn, Annie. "Michele Williams, Nov. Winner." Chris Kindness Award, November 30, 2022. https://chriskindnessaward.org/michele-williams-nov-winner/.

Han, Qijun, and Daniel R. Curtis. "Social Responses to Epidemics Depicted by Cinema." *Emerging Infectious Diseases* 26, no. 2 (2020): 389–94. http://doi.org/10.3201/eid2602.181022.

Harvard Graduate School of Education and Making Caring Common Project. "Loneliness in America: How the Pandemic Has Deepened an Epidemic of Loneliness and What We Can Do about It." February 2021. https://mcc.gse.harvard.edu/reports/loneliness-in-america.

Haslam, S. Alexander, Charlotte McMahon, Tegan Cruwys, Catherine Haslam, Jolanda Jetten, and Niklas K. Steffens. "Social Cure, What Social Cure? The Propensity to Underestimate the Importance of Social Factors for Health." *Social Science and Medicine* 198 (2018): 14–21. https://doi.org/10.1016/j.socscimed.2017.12.020.

Hawaiʻi Department of Land and Natural Resources. "Rare Plants." https://dlnr.hawaii.gov/ecosystems/rare-plants/.

Hawaii News Now Staff. "Hawaii Saw More than 10M Visitors Last Year, but Not Everyone Is Celebrating." January 30, 2020. https://www.hawaiinewsnow.com/2020/01/31/hawaii-saw-more-than-m-visitors-last-year-not-everyone-is-celebrating/.

Hiʻipaka LLC. "History of Waimea Valley." https://www.waimeavalley.net/about-waimea.

History.com. "Civilian Conservation Corps." Updated March 31, 2021. https://www.history.com/topics/great-depression/civilian-conservation-corps#section_6.

Holthaus, Eric. "The Firestorm Ravaging Northern California Cities, Explained." *Mother Jones*, October 10, 2017. https://www.motherjones.com/environment/2017/10/the-firestorm-ravaging-northern-california-cities-explained/.

Howard, Lisa. "Volunteering in Late Life May Protect the Brain against Cognitive Decline and Dementia." UC Davis Health News, July 20, 2023. https://health.ucdavis.edu/news/headlines/volunteering-in-late-life-

may-protect-the-brain-against-cognitive-decline-and-dementia
/2023/07.

Howarth, Josh. "Alarming Average Screen Time Statistics (2023)." *Exploding Topics* (blog), January 13, 2023. https://explodingtopics.com/blog/screen-time-stats.

How to PTA in Challenging Times: Texas PTA Rally Day 2021. National PTA, 2021. https://www.pta.org/docs/default-source/files/runyourpta/2021/how-we-pta/case-study---texas-pta-rally-day.pdf.

Jha, Alok. "Why Crying Babies Are So Hard to Ignore." *The Guardian*, October 17, 2012. https://www.theguardian.com/science/2012/oct/17/crying-babies-hard-ignore.

Jobs, Steve. "Steve Jobs Introducing the iPhone at MacWorld 2007." YouTube, January 9, 2007, video, 14:00. https://www.youtube.com/watch?v=x7qPAY9JqE4.

Johnson, M. T., J. M. Fratantoni, K. Tate, and A. S. Moran. "Parenting with a Kind Mind: Exploring Kindness as a Potentiator for Enhanced Brain Health." *Frontiers in Psychology* 13 (2022): 805748. https://doi.org/10.3389/fpsyg.2022.805748.

Jones, R. "It's Good to Give." *Nature Reviews Neuroscience* 7 (2006): 907. https://doi.org/10.1038/nrn2047.

Karlis, Nicole. "Before the Pandemic, They Were Introverts. Now They Aspire to Live More Extroverted Lives." *Salon.com*, June 6, 2021. https://www.salon.com/2021/06/06/introverts-post-pandemic/.

———. "How the 2010s Became the Decade of Self-Care." *Salon.com*, December 21, 2019. https://www.salon.com/2019/12/21/how-the-2010s-became-the-decade-of-self-care/.

———. "Lisa Luckett, 9/11 Widow, Explains How Tragedy Helps Us Grow." *Salon.com*, September 11, 2018. https://www.salon.com/2018/09/11/lisa-luckett-911-widow-explains-how-tragedy-helps-us-grow/.

———. "Why Doing Good Is Good for the Do-Gooder." *New York Times*, October 26, 2017. https://www.nytimes.com/2017/10/26/well/mind/why-doing-good-is-good-for-the-do-gooder.html.

———. "Why 'Social Distancing,' if Done Wrong, Can Make You More Vulnerable." *Salon.com*, March 15, 2020. https://www.salon.com/2020/03/15/why-social-distancing-if-done-wrong-can-make-you-more-vulnerable/.

Keltner, Dacher. "What's the Most Common Source of Awe?" *Greater Good Magazine*, January 24, 2023. https://greatergood.berkeley.edu/article/item/whats_the_most_common_source_of_awe.

Kim, Eric S., and Sara H. Konrath. "Volunteering Is Prospectively Associated with Health Care Use among Older Adults." *Social Science and Medicine* (January 2016): 122–29. https://doi.org/10.1016/j.socscimed.2015.11.043.

Kinnell, Galway. "Saint Francis and the Sow." In *Mortal Acts, Mortal Words*. Boston: Houghton Mifflin, 1980. Available at https://www.encyclopedia.com/arts/educational-magazines/saint-francis-and-sow.

Knight, Clarke A., Lysanna Anderson, M. Jane Bunting, et al. "Land Management Explains Major Trends in Forest Structure and Composition over the Last Millennium in California's Klamath Mountains." *Proceedings of the National Academy of Sciences* 119, no. 12 (March 14, 2022): e2116264119. https://doi.org/10.1073/pnas.2116264119.

Konrath, Sara H., Edward H. O'Brien, and Courtney Hsing. "Changes in Dispositional Empathy in American College Students Over Time: A Meta-Analysis." *Personality and Social Psychology Review* 15, no. 2 (May 2011): 180–98. https://doi.org/10.1177/1088868310377395.

Luckett, Lisa. "Love vs Fear. Can We See 9/11 in a New Light?" TEDxNew Bedford, December 7, 2016. https://www.youtube.com/watch?v=LOnMZXbII7M.

Luks, Allan. "Doing Good: Helper's High." *Psychology Today* 22, no. 10 (1988). Photocopy of article available at https://ellisarchive.org/sites/default/files/2019-10/Document_20191001_0008.pdf.

———. "Helper's High: The Healing Power of Helping Others." https://allanluks.com/helpers_high.

Lyubomirsky, Sonja. "On Studying Positive Emotions." *Prevention and Treatment* 3, no. 1 (March 7, 2000). https://psycnet.apa.org/doi/10.1037/1522-3736.3.1.35c.

Making Caring Common. "Relationship Mapping Strategy." Harvard University Graduate School of Education. Accessed November 11, 2023. https://mcc.gse.harvard.edu/resources-for-educators/relationship-mapping-strategy.

———. "Virtual Listening Deeply: Strategy and Lesson Plans." Harvard University Graduate School of Education. Accessed November 11, 2023. https://static1.squarespace.com/static/5b7c56e255b02c683659fe43

/t/5f2afe834450e97f1c74fcb2/1596653206496/Virtual+Listening+Deeply.pdf.

Maritime Museum of the Atlantic. "Halifax Explosion." Accessed October 18, 2023. https://maritimemuseum.novascotia.ca/what-see-do/halifax-explosion.

Marlin, B., M. Mitre, J. D'amour, et al. "Oxytocin Enables Maternal Behaviour by Balancing Cortical Inhibition." *Nature* 520 (2015): 499–504. https://doi.org/10.1038/nature14402.

Maui Nui Botanical Gardens. "Koa (Acacia koa)." Accessed August 19, 2023. https://mnbg.org/hawaiian-native-plant-collection/koa-acacia-koa.

May, Katherine. *Wintering: The Power of Rest and Retreat in Difficult Times*. Waterville, ME: Thorndike Press, 2021.

McAvoy, Audrey. "Agency to Hawaii Residents: Don't Hate on Tourists." Associated Press, April 24, 2016. https://www.seattletimes.com/nation-world/agency-to-hawaii-residents-dont-hate-on-tourists/.

McKinsey & Company. "Feeling Good: The Future of the $1.5 Trillion Wellness Market." April 8, 2021. https://www.mckinsey.com/industries/consumer-packaged-goods/our-insights/feeling-good-the-future-of-the-1-5-trillion-wellness-market.

Murphy, Chris. "Congressional Remarks from Senator Chris Murphy (D-CT)." YouTube, June 14, 2023, speech, 12:47. https://www.youtube.com/watch?v=SAPFkQfjNgI.

NASA Glenn Research Center. "Newton's Laws of Motion." Accessed November 11, 2023. https://www1.grc.nasa.gov/beginners-guide-to-aeronautics/newtons-laws-of-motion.

National Institute on Alcohol Abuse and Alcoholism (NIAAA). "Alcohol-Related Deaths, which Increased during the First Year of the COVID-19 Pandemic, Continued to Rise in 2021." April 12, 2023. https://www.niaaa.nih.gov/news-events/research-update/alcohol-related-deaths-which-increased-during-first-year-covid-19-pandemic-continued-rise-2021.

National Park Service. "Hurricane Andrew." Accessed October 18, 2023. https://www.nps.gov/articles/hurricane-andrew-1992.htm.

National Weather Service. "Hurricane Andrew: 30 Years Later." Accessed October 18, 2023. https://www.weather.gov/lmk/HurricaneAndrew30Years.

———. "Major Hurricane Maria." September 20, 2017. https://www.weather.gov/sju/maria2017.

"New Cigna Study Reveals Loneliness at Epidemic Levels in America: Research Puts Spotlight on the Impact of Loneliness in the U.S. and Potential Root Causes." Cigna, May 1, 2018. https://www.multivu.com/players/English/8294451-cigna-us-loneliness-survey/.

NHS England. "Social Prescribing: What Is Social Prescribing?" Accessed August 21, 2024. https://www.england.nhs.uk/personalisedcare/social-prescribing/.

NHS England. "Social Prescribing—The Power of Time and Connections." https://www.england.nhs.uk/personalisedcare/comprehensive-model/case-studies/social-prescribing-the-power-of-time-and-connections/.

9/11 Day. "Jay Winuk." https://911day.org/leaders/jay-winuk.

Oak Ridge Associated Universities. "Psychosocial Reactions: Phases of Disaster." Accessed March 5, 2024. https://www.orau.gov/rsb/pfaird/03-psychosocial-reactions-01-phases-of-disaster.html.

Online Etymology Dictionary. s.v. "volunteer." https://www.etymonline.com/word/volunteer.

Ortiz, Celina, and Jason Swinderman. "Eusocial and Colony Behavior in Ants." Reed College, Biology 342, Fall 2012. https://www.reed.edu/biology/courses/BIO342/2012_syllabus/2012_WEBSITES/COJS_animal-Behavior/index2.html.

Perrin, Andrew, and Sara Atske. "About Three-in-Ten U.S. Adults Say They Are 'Almost Constantly' Online." Pew Research Center, March 26, 2021. https://www.pewresearch.org.

Phillips, Adam, and Barbara Taylor. *On Kindness.* New York: Picador, Macmillan, 2020.

Physiopedia. s.v. "vagus nerve." https://www.physio-pedia.com/Vagus_Nerve.

Polyvagal Institute. "What Is the Polyvagal Theory?" https://www.polyvagalinstitute.org/whatispolyvagaltheory.

Porges, S. W., and S. A. Furman. "The Early Development of the Autonomic Nervous System Provides a Neural Platform for Social Behavior: A Polyvagal Perspective." *Infant and Child Development* 20, no. 1 (January/February 2011): 106–18. https://doi.org/10.1002/icd.688.

Portes, Alejandro. "Social Capital: Its Origins and Applications in Modern Sociology." *Annual Review of Sociology* 24 (1998): 1–24.

Prince, Samuel Henry. "Catastrophe and Social Change: Based upon a Sociological Study of the Halifax Disaster." PhD diss., Columbia University, New York, 1920. https://www.gutenberg.org/files/37580/37580-h/37580-h.htm.

Putnam, Robert D. "Bowling Alone: America's Declining Social Capital." *Journal of Democracy* 6, no. 1 (January 1995): 65–78. https://www.journalofdemocracy.org/articles/bowling-alone-americas-declining-social-capital/.

Quarantelli, E. L. "Disaster Studies: An Analysis of the Social Historical Factors Affecting the Development of Research in the Area." *International Journal of Mass Emergencies and Disasters* 5 (1987): 285–310. http://udspace.udel.edu/handle/19716/1335.

Reese, Hope. "How a Bit of Awe Can Improve Your Health." *New York Times*, January 3, 2023. https://www.nytimes.com/2023/01/03/well/live/awe-wonder-dacher-keltner.html.

Refinery29. "I Live in Los Angeles, Make $75,000 a Year, and Spent $2,003 on My Wellness Routine This Week." *Feel Good Diaries*, June 17, 2019. https://www.refinery29.com/en-us/2019/06/235357/wellness-routine-barrys-bootcamp-smoothies.

Robertson, C. E., N. Pröllochs, K. Schwarzenegger, et al. "Negativity Drives Online News Consumption." *Nature Human Behaviour* 7 (2023): 812–22. https://doi.org/10.1038/s41562-023-01538-4.

Russell, Dan, Letitia Anne Peplau, and Mary Lund Ferguson. "Developing a Measure of Loneliness." *Journal of Personality Assessment* 42, no. 3 (July 1978): 290–94. http://dx.doi.org/10.1207/s15327752jpa4203_11.

Ryan, R. M., and E. L. Deci. "Self-Determination Theory and the Facilitation of Intrinsic Motivation, Social Development, and Well-Being." *American Psychologist* 55 (2000): 68–78.

Schier-Akamelu, Rebecca. "2023 Caregiver Burnout and Stress Statistics." A Place for Mom, last updated June 13, 2023. https://www.aplaceformom.com/senior-living-data/articles/caregiver-burnout-statistics.

Screen Time Action Network. Letter to Jessica Henderson Daniel, PhD, ABPP, President, American Psychological Association. August 8, 2018.

https://screentimenetwork.org/apa?eType=EmailBlastContent&eId=502
6ccf8-74e2-4f10-bc0e-d83dc030c894.

Simkhovitch, Vladimir G. "Mutual Aid a Factor of Evolution, by P. Kropotkin." *Political Science Quarterly* 18, no. 4 (December 1903): 702–5. https://doi.org/10.2307/2140787.

Solnit, Rebecca. *A Paradise Built in Hell: The Extraordinary Communities That Arise in Disaster*. New York: Penguin, 2010.

Stambor, Zak. "A Key to Happiness." *Monitor on Psychology* 37, no. 9 (October 2006). https://www.apa.org/monitor/oct06/key.

Statt, Nick. "The Creators of the iPhone Are Worried We're Too Addicted to Technology." *The Verge*, June 29, 2017. https://www.theverge.com/2017/6/29/15893960/apple-iphone-creators-smartphone-addiction-ideo-interview.

Straub, Adam M., Benjamin J. Gray, Liesel Ashley Ritchie, and Duane A. Gill. "Cultivating Disaster Resilience in Rural Oklahoma: Community Disenfranchisement and Relational Aspects of Social Capital." *Journal of Rural Studies* 73 (January 2020): 105–13.

Sturm, R., and D. A. Cohen. "Free Time and Physical Activity among Americans 15 Years or Older: Cross-Sectional Analysis of the American Time Use Survey." *Preventing Chronic Disease* 16 (September 26, 2019). http://dx.doi.org/10.5888/pcd16.190017.

Sweetland Edwards, Haley. "You're Addicted to Your Smartphone. This Company Thinks It Can Change That." *Time*, April 12, 2018. https://time.com/5237434/youre-addicted-to-your-smartphone-this-company-thinks-it-can-change-that/.

Taylor, Alan. "Photos: The Volunteers." *The Atlantic*, April 2, 2020. https://www.theatlantic.com/photo/2020/04/photos-the-volunteers/609149/.

Testard, Camille, Sam M. Larson, Marina M. Watowich, Noah Snyder-Mackler, Michael L. Platt, and Lauren J. N. Brent. "Rhesus Macaques Build New Social Connections after a Natural Disaster." *Current Biology* 31, no. 11 (2021): 2299–2309. https://doi.org/10.1016/j.cub.2021.03.029.

Thai Life Channel. "Unsung Hero." YouTube, April 3, 2014, commercial, 3:05. https://www.youtube.com/watch?v=uaWA2GbcnJU.

Thompson, Derek. "Click Here if You Want to Be Sad." *The Atlantic*, March 24, 2023. https://www.theatlantic.com/newsletters/archive/2023/03/negativity-bias-online-news-consumption/673499/.

Thoreau, Henry David. *The Journal of Henry David Thoreau, 1837–1861*. Edited by Damion Searls. New York: New York Review Books Classics, 2009.

Trivers, Robert L. "The Evolution of Reciprocal Altruism." *Quarterly Review of Biology* 46, no. 1 (March 1971): 35–57. http://www.jstor.org/stable/2822435?origin=JSTOR-pdf.

Tucker, Jill. "Santa Rosa Schools Reopen after Fires, Ready to Help Students with Stress." *San Francisco Chronicle*, October 27, 2017. https://www.sfchronicle.com/education/article/Wildfire-danger-is-past-but-stress-can-linger-12312685.php.

University of Pennsylvania. "How Do Natural Disasters Shape the Behavior and Social Networks of Rhesus Macaques?" Press release, April 8, 2021. https://penntoday.upenn.edu/news/Penn-neuroscience-natural-disasters-behavior-social-networks-rhesus-macaques.

University of Wisconsin–Madison. "Brain Can Be Trained in Compassion, Study Shows." Press release, May 22, 2013. https://news.wisc.edu/brain-can-be-trained-in-compassion-study-shows/.

U.S. Census Bureau. "New Data Reveal Most Populous Cities Experienced Some of the Largest Decreases." May 26, 2022. https://www.census.gov/library/stories/2022/05/population-shifts-in-cities-and-towns-one-year-into-pandemic.html.

———. *Remembering Pearl Harbor: A Pearl Harbor Fact Sheet*. https://www.census.gov/history/pdf/pearl-harbor-fact-sheet-1.pdf.

U.S. Forest Service. "First Returners." Accessed October 18, 2023. https://www.fs.usda.gov/Internet/FSE_DOCUMENTS/fseprd575963.pdf.

U.S. Surgeon General. *Our Epidemic of Loneliness and Isolation: The U.S. Surgeon General's Advisory on the Healing Effects of Social Connection and Community*. 2023. https://www.hhs.gov/sites/default/files/surgeon-general-social-connection-advisory.pdf.

Walker, J., A. Kumar, and T. Gilovich. "Cultivating Gratitude and Giving through Experiential Consumption." *Emotion* 16, no. 8 (2016): 1126–36. https://doi.org/10.1037/emo0000242.

Warneken, Felix, and Michael Tomasello. "Altruistic Helping in Human Infants and Young Chimpanzees." *Science* 311 (2006): 1301. https://doi.org/10.1126/science.1121448.

Wenner Moyer, Melinda. "Kids as Young as 8 Are Using Social Media More than Ever, Study Finds." *New York Times*, March 24, 2022. https://www.nytimes.com/2022/03/24/well/family/child-social-media-use.html.

Williams, George. *Adaptation and Natural Selection*. Princeton, NJ: Princeton University Press, 1966.

WKYC Channel 3. "Boys Do Read: 8-Year-Old Boy Starts Reading Movement in Cleveland." November 23, 2018. https://www.youtube.com/watch?v=aJbTRY2xwfY.

Wolf, Jessica. "Is Kindness Contagious?" *UCLA Magazine*, January 5, 2023. https://newsroom.ucla.edu/magazine/bedari-kindness-institute-contagious.

Yotopoulos, Amy. "Three Reasons Why People Don't Volunteer, and What Can Be Done about It." Stanford Center on Longevity. https://longevity.stanford.edu/three-reasons-why-people-dont-volunteer-and-what-can-be-done-about-it/.

Index

abandonment, sense of, 30
absenteeism crisis in school, 149–50, 152
abundance, sense of, 164–65, 171
adaptation, 26
Adaptation and Natural Selection (Williams), 27
adaptation-level theory, 54
aging: and loneliness, 47; and stress, 85
Alberti, Fay Bound, 42
Aldrich, 16, 18, 30–31
altruism, acts of, 5–6, 20, 22; and brain, 86–93. *See also* kindness; well-being
Alzheimer's Association International Conference, 112–13
Amster, Michael, Dr., 173–75
anger, 30, 65, 154–55
antisocial behavior, 91
autonomic nervous system, 50–51, 58–59, 141–42, 143, 158
AWE (Attention, Waiting, Exhaling/ Expanding), 174, 189
awe: in caregiving, 174–75; in witnessing kindness, 172–73

Awe: The New Science of Everyday Wonder and How It Can Transform Your Life (Keltner), 172

Baek, Elisa, 47–48
Bartalotti, Carlo, 11–12
Bedari Kindness Institute, 172
behavioral prescriptions, 160–61
Be KIND (Kin-Initiated Nice Deeds) initiative, 73–74
belonging, sense of, 129–34
bias, 59, 83, 143, 169, 175, 178
Biography of Loneliness: The History of an Emotion, A (Alberti), 42
boredom, 135
bounded solidarity, 20–25, 84–85. *See also* caring; community; social connections
Boundless Mind, 40
Bowling Alone: The Collapse and Revival of American Community (Putnam), 33, 57, 117
Boys Do Read, 74
brain agility: and social support, 85; and volunteering, 112–13

[219]

brain and altruism, 86-93; training the brain, 89-91
BrainHealth Project, 86
Brickman, Philip, 54, 55, 56
Brown, Ramsay, 40
burnout and caregiving, 173-75

Cacioppo, John, 41-43, 46-47, 60
California Climate Action Corps, 119-20, 122
California Service Corps, 119-20, 122-25
Callaghan, William M., 100
Campbell, Donald T., 54, 56
cancer, 138-39
caregiving, 141; and burnout, 173-75
caring: in crisis, 5-6, 10-13; for others, 149-54; for others, teaching, 156-57. *See also* bounded solidarity
Carpenter, Clarence, 82
Carter, Sue, Dr., 158-59
Chase, David, 32-33
children: prosocial behaviors of, 29, 73, 92; teaching kindness to, 73-75, 91-92, 169. *See also* students
Chris Kindness Awards, 168-69
Christie, Greg, 37
chronic diseases and loneliness, 137-38
City on the Edge: The Transformation of Miami (Portes), 21
civic engagement, 33-35, 81, 117
Civilian Conservation Corps (CCC), 122
CLICK (Connect, Listen, Investigate, Communicate kindness, and Keep in touch), 127-28, 129-30, 131-32

Climate Action Corps, 124
cognitive behavioral therapy (CBT), 67-68
cognitive reappraisal, 67-68, 90
Cole, Steve, Dr., 38-39, 40, 136-44, 174
collective action, 111, 175
collective behavior, 20
College Corps, 119, 120
Color a Smile, 75
Color Me a Rainbow, 74
common purpose, sharing, 109-10
Common Sense Media, 40
community: of mutual benefit, 143; service, 118, 125. *See also* bounded solidarity; social connections
compassion meditation, 89-90
compassion training, 91-92
Connections at the Capitol, 154
Cordova, Robbie, 123-24
coregulation fix, 49-60
Cotta, Cassandra, 14-15
COVID-19 pandemic, 16-17, 44, 91-92; news during, 176; and social isolation, 140-41; volunteering during, 12
Cregg, David, 66-70
crisis: absenteeism crisis in school, 149; caring in, 5-6, 10-13; "Disillusionment Phase" of, 29-30; "Heroic Phase" of, 29; "Honeymoon Phase" of, 29; "Recovery Phase" of, 29-30; self-interest during, 16-17; social bonds during, 26-31; and social connections, 82-84; social order breakdown during, 16-17; students, loneliness crisis among, 127-34. *See also* loneliness

culture of caring, 8, 9, 151, 178
curiosity and kindness, 86

A Daily Dose of Reading, 73
Darwin, Charles, 26
declarative learning, 91
deep listening, 152–54
Descent of Man and Selection in Relation to Sex, The (Darwin), 26
Devlin, Sean, 176–78
Dillon, Michele, 135
disenfranchisement, 25
"Disillusionment Phase" of a crisis, 29–30
dissociation and atomization, 142
dopamine, 87, 88
dorsolateral prefrontal cortex, 91
Doty, James, Dr., 39
durable solidarity, 21–22, 85. *See also* bounded solidarity

elderly people: loneliness in, 44–45; volunteering among, 113–14, 121, 189
empathy, 91; and altruism, 155–56; building, 156–57; in listening, 152–54, 155–56
endorphins, 112
epigenetic clock, 114
executive function and volunteering, 113
existential vacuum, 135

Family Kindness Festival, 74–75
fear, 8, 17, 28, 39, 58, 87, 88, 159
feedback loop, 43, 46, 51, 59, 85
Feel Good Diaries blog, 53
foreign exchange students, sense of belonging among, 134

Frankl, Viktor, 134–35
Fratantoni, Julie, Dr., 85–86, 91–92
Fredrickson, Barbara, 65
free time, volunteering, 37, 39, 111, 154
Frontiers in Psychology (journal), 91
Fryday, Josh, 80–81, 117–22, 124–25
functional magnetic resonance imaging (fMRI), 90–91

Garcia, Carmen, 169
gene expression, 138, 140, 141–42, 144
generosity, acts of, 164–66
Getz, Lowell, 159
ghrelin hormone, 46
global wellness market, 55
Good Bones (poem, Smith), 170–71
Grant, Don, Dr., 37–38
group volunteering, 9, 115; for Mālama Hawaiʻi, 102–12; and resilience, 112

Hamilton, William D., 26–27
happiness interventions, 65
"happy hormone" (dopamine), 88
Haslam, Catherine, 56–58
Hawaiʻi Tourism Authority (HTA), 94–95
healing, 12, 53, 64, 96–97, 152
hedonic treadmill theory, 54–56
helper's high, 112
Hero Foundation, 10–12
"Heroic Phase" of a crisis, 29
Herrera, Michael, 103, 106, 110
higher-order thinking, 86
Hiʻipaka LLC, 103
"Honeymoon Phase" of a crisis, 29
hope, 60, 67, 178

human capital, 24
hyperindividualism, 153–54

IHD Longitudinal Study, 135
inclusive fitness, 27
individualism, 8, 22, 32, 153–54
Infant and Child Development: Prenatal, Childhood, Adolescence, Emerging Adulthood (journal), 58
inferior parietal cortex, 91
In the Course of a Lifetime (Dillon), 135

Journal of Henry David Thoreau, 1837–1861, The (Thoreau), 42
Junger, Sebastian, 5

Kaʻanāʻanā, Kalani, 94–97
Kapaemahu, 96–97
karma, 170–71
Keltner, Dacher, 172–73
Kim, Eric, Dr., 113–14, 121
Kind Minds with Moozie program, 92
kindness, 63–65; acts of, 66–72, 89, 114–15, 164–66, 168–70, 176, 186; as contagious, 75, 170, 172, 177–78; and resilience, 91–92; timing of, 66; training, 91–92; witnessing, 172–73
Kinnell, Galway, 59
kin selection, 27
Klein, Naomi, 22
Konrath, Sara, 156
Kropotkin, Peter, 16
Kumar, Amit, 70–72

Latimer, McKenzie, 104–9, 111–12
Lawrence, Crystal (volunteer), 80
Leary, Jonathan, 51–52, 55, 56

link worker, 76–77
listening: empathy in, 152–54, 155–56; practices, 152–54
logotherapy, 134
loneliness: and chronic diseases, 137–38; crisis among students, 127–34; in elderly people, 44–45; as epidemic, 35–36; and physical health, 46–47; and political divisiveness, 154–55; and premature aging, 47; and psychiatric disorders, 45; and PTSD, 138; score, 43–44; and social media, 44–45, 54; as social pain, 41–42; vs. solitude, 42; and stress, 138; and technology, 54
Loneliness: Human Nature and the Need for Social Connection (Cacioppo), 41
Lono (Hawaiian god), 103
Lor, Yi, 113
Luks, Allan, 112
Lyubomirsky, Sonja, 65–66

Making Caring Common project, 150–54
mālama mindset, in Hawaii, 94, 96, 98–99
Manning, Glenn, 149–51, 154, 156
Man's Search for Meaning (Frankl), 134–35
Markle, Elizabeth, Dr., 160–66
material possessions and happiness, 70–71
May, Katherine, 4–5, 27
McGroarty, Beth, 54
meaning, sense of, 57, 134–35
Merchant, Brian, 37
mesolimbic reward system, 87

mobility data, 30–31
Moll, Jorge, 86–87
monocytes, 138–39
Moore, Cat, 126–34
Morgenstern, Joan, 72–76, 145
Morse, Dan, 75–76, 77–78
Murphy, Chris, 154–55
Murthy, Vivek, Dr., 35–36, 45
Mutual Aid: A Factor of Evolution (Kropotkin), 16
mutual aid networks, 15–16, 24
mutual benefit, community of, 143

National Health Service (NHS), England, 76–77
National Opinion Research Center (NORC), 18
natural selection theory, 26
Nature Human Behaviour (journal), 175
Navarro, Crystal, 123
negative emotions, 65
negative feedback loop, 43, 46
negative news and mental health, 175–77
negativity bias, 175
"neighbor to neighbor" program, 121
neuroplasticity, 89
news, stress caused by, 175–77
Newsom, Gavin, 118
Nice News, 177–78

One Nation after Trump (Dionne et al.), 35
On Kindness (Phillips), 72
Open Source Wellness, 161–66
Ording, Bas, 37
oxytocin, functioning of, 158–60

Paine, David, 79–80
Paradise Built in Hell: The Extraordinary Communities That Arise in Disaster (Solnit), 5
personal well-being and money, 70–71
persuasive technology, 40
Phillips, Adam, 72
Platt, Michael, 82–85
PLOS One (journal), 176
political divisiveness, loneliness and, 154–55
polyvagal theory, 50–51
Porges, Stephen, 50–51, 58–59
Portes, Alejandro, 20–21
positive emotions, 65, 88, 91, 182
positive feedback loop, 51
positive news, 177–78
Power of Awe, The (Amster), 173
prefrontal cortex, 86
premature aging, 47
preparedness, individual, 15–16
procedural learning, 91
prosocial behaviors, 18, 25, 71, 73; of children, 29, 73, 92
prosocial contagion, 172
psychiatric disorders and loneliness, 45
psychological barriers, 72
purpose, sense of, 135–36, 143, 188
Putnam, Robert, 33–35, 57, 117

reciprocal altruism, 188
reciprocity, 23, 34
"Recovery Phase" of a crisis, 29–30
Redistribution Game, 90
regenerative tourism, 95–96
Reimagining Citizenship, 118
relational culture, 127

relationship mapping, 151–52
religion: and altruism, 20–21; downfall of, 144, 157; and fear, 58; and sense of purpose, 135–36
resilience: and group volunteering, 112; and kindness, 91–92; vs. recovery, 30–31; social support enabling, 84–86
reward pathway, 87
RNA profiles of lonely people, 137–38
Roosevelt, Franklin D., 122
Ross, Alan, 167–70
Rousseau, 25
Rupprecht, Michael, 4–5, 11–12

scan method, 83–84
screen time, 36–40
self-care, 53–54, 163, 188; caring of others as, 186, 187; definitions of, 163; modern day, 53–54; and sense of purpose, 188
self-determination theory, 69
self-serving survival, 59
Senders, Shelly, Dr., 73, 74, 78
Senders Pediatrics, 73, 77
service to commemorate 9/11, 78–81
shared meaning, 48
shared values, 142
Shock Doctrine, The (Klein), 22
Smith, Derrick, 74
Smith, Maggie, 170–71
social anxiety, 66, 91
social bonding effect, 159
social bonds during crisis, 26–31
social capital, 20, 34
social change and catastrophe, 20

social connections: as basic psychological need, 69; bridge, 129–30; building, 76; and crisis, 82–84; and nervous system, 143; and physical health, 46; and survival, 18–19. *See also* bounded solidarity; community
social contract, 25
social cure, 56–58
social isolation, 16, 43, 44–45, 140–41, 166
social media, 156; and loneliness, 44–45, 54
social monogamy, 159
social order breakdown during crisis, 16–17
Social Prescribing USA, 76
social prescription, 76–77
social support: enabling resilience, 85–93; as response to stressors, 82–85
social wellness club, 52–53
societal health and loneliness, 45
society's response to a disaster, 17–18
solitude vs. loneliness, 42
Solnit, Rebecca, 5, 22
Spencer, Herbert, 26
spiritual practice, 163
spiritual seekers, 135
Straub, Adam, 22–25
stress: and brain damage, 86; and loneliness, 138; and social support, 85
students: loneliness crisis among, 127–34. *See also* children
Supplemental Nutrition Assistance Program (SNAP), 120

sympathetic nervous system, 86
systemic caring, 125. *See* empathy

Taylor, Barbara, 72
technology and loneliness, 38–39, 54
Thomas, Emiliana Simon, Dr., 188
Thompson, Derek, 175
Thoreau, Henry David, 42
togetherness, sense of, 5
Tomasello, Michael, 29
"Tree Army," 122
Tribe: On Homecoming and Belonging (Junger), 5
Trivedi, Jennifer, 18
Trivers, Robert, 187–88
Tronick, Edward, 50
trust building, 16

UCLA Loneliness Scale, 43–44, 47
Upworthy (website), 175
USS *Missouri*, volunteerism on, 98–101

vagus nerve, 49–50, 51, 59, 174
vasopressin, functioning of, 159–60
ventral striatum, 88, 143
verbal episodic memory and volunteering, 113
virtuous reflex, 141
volunteer, the term, 116–17
volunteering: barriers to, 120–21; and brain agility, 112–13; creating culture of, 118–19, 122–25; and older adults, 113–14, 121
volunteerism: for restoration of koa trees, 102–12; on USS *Missouri*, 98–101. *See also* mālama mindset, in Hawaii
vulnerability, 16, 43, 72, 109, 131, 164

Waimea Valley: history, 102–3; restoration of koa trees, 102–9; volunteerism for restoration of koa trees, 102–12
Warneken, Felix, 29
Weissbourd, Richard, 44, 150–52, 154–55
well-being: and acts of kindness, 65–66, 69, 73, 89, 114–15, 176, 186; mental, 65, 70, 124; of others, 5–6, 135, 161; and self-care, 163; sense of, 72; and workplace culture, 140
wellness-focused activities, 52–54, 56–59
wellness market, global, 55
Willard, Dallas, 126
Williams, George, 27
Williams, Michele, 168–69
Williamson, Keven, 98–99
Wintering (May), 4, 27
Winuk, Jay, 79–80
witnessing kindness, effect of, 172
workplace culture and well-being, 140

Youth Jobs Corps, 119, 123

Founded in 1893,
UNIVERSITY OF CALIFORNIA PRESS
publishes bold, progressive books and journals
on topics in the arts, humanities, social sciences,
and natural sciences—with a focus on social
justice issues—that inspire thought and action
among readers worldwide.

The UC PRESS FOUNDATION
raises funds to uphold the press's vital role
as an independent, nonprofit publisher, and
receives philanthropic support from a wide
range of individuals and institutions—and from
committed readers like you. To learn more, visit
ucpress.edu/supportus.